Astronomers' Observing Guides

For other titles published in this series, go to
www.springer.com/series/5338

Roger Dymock

Asteroids and Dwarf Planets
and How to Observe Them

with 152 Illustrations

 Springer

Roger Dymock
67 Haslar Crescent
Waterlooville
Hampshire
PO7 6DD
United Kingdom
roger.dymock@ntlworld.com

Series Editor
Dr. Mike Inglis, BSc, MSc, Ph.D.
Fellow of the Royal Astronomical Society
Suffolk County Community College
New York, USA
inglism@sunysuffolk.edu

Please note that additional material for this book can be downloaded from http://extras.springer.com

ISBN 978-1-4419-6438-0   e-ISBN 978-1-4419-6439-7
DOI 10.1007/978-1-4419-6439-7
Springer New York Dordrecht Heidelberg London

Library of Congress Control Number: 2010938909

Printed on acid-free paper

Springer is part of Springer Science+Business Media (www.springer.com)

Acknowledgements

While writing this book I requested the help of many amateur and professional astronomers with images, diagrams, specific projects, and those sections with which I was not familiar. I have been truly amazed with their responses – my requests were answered speedily and additional images and information were frequently offered. My only hope is that this book does justice to their efforts.

First of all I must thank my wife, Jean for her support. As long as I don't wake her in the middle of the night or interfere with her tennis she is happy for me to indulge in matters astronomical. Actually she does have more than a passing interest in astronomy, having completed the same basic course as I did, run by Hampshire Astronomical Group's Past President, Robin Gorman. Both having had a painful bunion removed I sometimes felt that their conversation was more anatomical than astronomical! When I first mentioned writing this book she had visions of Harry Potter like riches – well you never know, but I definitely (well, almost definitely) won't be writing another!

Many, many thanks to the following (apologies for not including titles as they were not always given in communications): David C. Agle (National Aeronautics and Space Administration Jet Propulsion Laboratory – NASA JPL), Paul G. Allen Charitable Foundation, Eamonn Ansbro, Molly Birtwhistle, Peter Birtwhistle, John Broughton – Scanalyzer and ScanTracker, Michael R. Buckley (NASA Johns Hopkins University Applied Physics Laboratory), Marc Buie (Lowell Observatory), L. Calçada (European Southern Observatory – ESO), California Institute of Technology – Caltech, Alan Chamberlain, Michael Clarke, Steven K. Croft, Martin Crow, Earth Impact Database, ELB software – Megastar, Andrew Elliott, Brian Fessler (Lunar and Planetary Institute – LPI), Eric Frappa (Euraster), Maurice Gavin, Eric Graff, Mary Ann Hager (LPI), Alan W. Harris (Space Science Institute), Tsutomu Hayamizu, Carl Hergenrother, David Higgins, Nadia Imbert-Vier (European Space Agency – ESA), Chris Hooker, Suzanne H. Jacoby (Large Synoptic Survey Telescope), Nick James, David Jewitt, Richard Judd (Hampshire Astronomical Group), Scott Cardel (Caltech, Palomar Observatory), Mikko Kaasalainen (Tampere University of Technology), Guido Kosters (ESA), David Kring (LPI), Stefan Kurti, Marco Langbroek, Rob Matson, Robert McMillan (Spacewatch FMO Project), Rob McNaught (Siding Spring Observatory), Richard Miles (Director, Asteroids and Remote Planets Section, British Astronomical Association), Martin Mobberley, NASA National Space Science Data Center – NSSDC, NASA Skymorph, Near Earth Asteroid Rendezvous mission – NEAR, Tomomi Niizeki (Japan Aerospace Exploration Agency – JAXA), Richard N. Nugent, Chris Peterson, Project Pluto – Find_Orb and Guide, Herbert Raab – Astrometrica, Monty Robson, Lou Scheffer, Jean Schwaenen (European Asteroidal Occultation Network – EAON), Brett Simison (Panoramic Survey Telescope & Rapid Response System – Pan-STARRS), John Saxton, Jean Schwaenen (EAON), Space Telescope Science Institute – STScI, Tim Spahr (Minor Planet Center – MPC), Alan Stern (Southwest Research

Institute), John Sussenbach, Roy Tucker, University of Arizona, Višnjan School of Astronomy, Brian Warner (Minor Planet Observer – MPO and Collaborative Asteroid Lightcurve Link – CALL), Gareth V. Williams (MPC), Rich Williams (Sierra Stars Observatory Network – SSON).

My astronomical 'career' began with my local astronomical society – the Hampshire Astronomical Group – and I subsequently joined the British Astronomical Association, *The Astronomer* Group, and the Royal Astronomical Society. Many members of those organizations have given me advice along the way, for which I am grateful.

All my (electronic) scribblings have been proofread by Hazel McGee, long time BAA member and editor of the *Journal* of the British Astronomical Association. We haven't always seen eye-to-eye, but mostly we did, and thus the text is somewhat improved from the original. Thank you so much Hazel.

Finally my thanks to my Springer contacts: Michael Inglis, Turpana Molina, Maury Solomon, and John Watson. Their requirement to submit a detailed structure of the book helped me break down its writing into manageable sections and keep to the proposed schedule. Martin Mobberley likened writing a book to climbing a mountain and having those 'camps' along the way certainly eased the path to the summit.

It seemed quite an omission to me that, to the best of my knowledge, no similar book existed, since many other aspects of amateur astronomy, e.g., comets, lunar, planetary, solar and variable stars were well catered for. Our knowledge of asteroids has increased rapidly of late – even during the time it has taken to write this book. Perhaps having set the ball rolling others will be encouraged to follow in my footsteps.

About the Author

Roger Dymock lives in Hampshire, England. He is a Fellow of the Royal Astronomical Society and was the Director of the Asteroids and Remote Planets Section of the British Astronomical Association from 2005 to 2008. His published work includes *Journal of the BAA*: "The Observapod – a GRP observatory"; Minor Planet Bulletin, No. 32 2005: "Lightcurve of 423 Diotima"; *Sky at Night* magazine: "How to track an asteroid"; and *Journal of the BAA*: "A method for determining the V magnitude of asteroids from CCD images" (jointly with Dr Richard Miles).

Contents

Contents

Part I

Asteroids and Dwarf Planets

Introduction

The asteroid world is ever changing. In no other area of astronomy are new objects being discovered at so fast a rate. New theories as to the evolution of asteroids, particularly beyond Neptune, are continually being put forward. Automated telescopic surveys and space missions are constantly making new discoveries and providing new data as to the nature of asteroids and dwarf planets. All this will be covered in the first part of the book. There is much the amateur, even with modest equipment, can do, as will be explained in Part II.

There is a downside to this rate of change. There are sometimes competing theories which, at the time of writing, have yet to be resolved, and it may be that some well established ideas are overturned in the near future – how planets have captured asteroids and turned them into moons, for example. Nothing new in this, but be aware that something you may read here is not necessarily wrong but may just be out of date or one of several theories on the subject.

After our Sun and its retinue of planets had formed around 4.5 billion years ago, there was a fair amount of 'builders rubble' left over. These lumps of material, stony, iron, or a mixture of both, some solid, and some loosely bound collections of smaller pieces could be found in large numbers in what is now known as the Main Belt between Mars and Jupiter, and we refer to them as asteroids. They do, of course, turn up in many other places with names (and orbits) that are by no means constant over the years: Vulcanoids, Trojans, Centaurs, Edgeworth–Kuiper Belt Objects, Trans-Neptunian Objects, Plutinos, Plutoids, Scattered Disk Objects, the Oort Cloud. You will find more on these groups in Chap. 3, and how they got to be where they are in Chap. 5. It is not only our planetary system that has such bodies. The search for extrasolar planets has turned up disks around other stars that may well contain asteroid belts similar to our own.

To return to our own locale, here is a short story. (Beware that this is a brief, generalized description of how an asteroid might journey from the Main Belt to the inner Solar System and may not be specific to this particular object. A fuller description is given in Chap. 5.)

Sunlight falling on a particular rotating asteroid exerted a force that caused it to slowly spiral outwards from its original location in the Main Belt. After several millions of years it reached an unstable area devoid of any of its companions – a Kirkwood Gap. Here it came under the gravitational influence of the gas giant planet Jupiter, which caused the eccentricity of its orbit to change significantly over a few tens of thousands of years. This change in eccentricity caused the asteroid to become a Mars-crosser and then to arrive in the vicinity of Earth – a near-Earth

R. Dymock, *Asteroids and Dwarf Planets and How to Observe Them*,
Astronomers' Observing Guides, DOI 10.1007/978-1-4419-6439-7_1,
© Springer Science+Business Media, LLC 2010

object, or NEO. Further planetary perturbations circularized the orbit to ensure it remained in the vicinity of Earth. This particular asteroid, Apophis, is not only an NEO but also an NVO, a near-Venus object.

Its arrival was first noted in June 2004 by astronomers using the Steward Observatory's telescope on Kitt Peak, Arizona, and a provisional designation, 2004 MN_4, was assigned. It was not seen again until the following December, when it was rediscovered by the Siding Spring Survey in Australia. Orbital calculations from these two sets of observations, by NASA's Jet Propulsion Laboratory and the University of Pisa, suggested that there was a greater than 1% chance of an impact on Earth in 2029, causing damage over a large area. This merited a rating of 4 on the Torino scale (explained in Chap.6), used by astronomers to define both the chances of an impact and the resulting devastation to life on Earth.

Not since the scale was formulated in 1999 had such a high rating been assigned to an incoming asteroid. By projecting the orbit backwards in time astronomers found the object on images obtained by the Spacewatch telescope on Kitt Peak in March 2004 (a pre-covery). Amateur astronomers also played a part in helping to define the orbit of 2004 MN_4 more accurately. Both the Goodricke–Pigott Observatory in the United States and the Observatori Astronomic de Mallorca submitted astrometry to the Minor Planet Center. Radar observations in January 2005 using the Arecibo radio telescope enabled the orbit to be further refined and showed that there was now no chance of an impact on the original date, but there would be a very close pass in 2029 and the possibility of an impact in 2036. Figure 1.1 shows the error ellipse, or region of uncertainty, through which 2004 MN_4 was predicted to pass when close to Earth. As can be seen Earth was a possible target until more accurate data was obtained and a new orbit calculated.

Without knowing the size or composition of an asteroid the damage likely to be caused by an impact is hard to assess. Spectra obtained in January 2005 showed that this Aten class asteroid, now numbered and named (99942) Apophis, was made of material similar to an ordinary chondritic meteorite. This data enabled the reflectivity, or albedo, and thus the size of the body to be calculated – approximately 270 m in diameter.

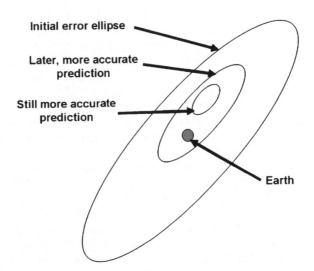

Initial error ellipse

Later, more accurate prediction

Still more accurate prediction

Earth

Fig. 1.1. Error ellipse, or region of uncertainty, as 2004 MN4 approached Earth (Credit Lou Scheffer).

Earth's gravity, being much weaker than that of Jupiter, does not usually affect the orbits of asteroids to such a great extent, but the proximity of Apophis to Earth in 2029 will cause a significant change in that asteroid's orbit. Exactly what that change will be cannot be calculated with any precision until the close approach. Amateur astronomers, with relatively modest equipment, will be able to play an important part along with professional optical and radar observers. To calculate an orbit you need accurate astrometry (measures of the asteroid's position) – how to do this is explained in Chaps.10 and 11. At its closest Apophis will reach third magnitude and will thus be one of the very few asteroids visible to the naked eye. Although a bonus to those without binoculars or telescopes, it is not one which we should hope will occur too often! In fact, at its closest, Apophis may be too bright for the sensitive detectors used by the automated surveys, thus giving amateurs an even more significant role.

Is there a point to the above story? Yes – more than one. It both introduces terminology that will be explored in greater depth in this book, and touches on the part that amateurs can still play in improving our understanding of these bodies once labeled the 'vermin of the skies.'

This book is aimed at those who can find their way around the sky and have a general knowledge of matters astronomical. You will probably have been observing for a couple of years or so and have access to the required equipment – your own, your local astronomical society, or robotic telescopes. Note that, for brevity, the term 'asteroids' will be used to encompass both those bodies and dwarf planets. 'Asteroids' is certainly easier on the mind than 'Small Solar System Bodies with the exception of Comets,' but more on that in Chap.2! Beginners are not ignored. Chap.8 will describe what is necessary to start you down the road to enjoying simply finding asteroids ("star-like objects," for that is what the word means) among the stars.

As you will find in Part II of this book, you can observe asteroids with a wide range of instruments and imagers: binoculars, refracting and reflecting telescopes, webcams, digital SLR cameras, video cameras, and CCD imagers. You can brave the elements and sit outside with your own equipment, or operate it remotely from the comfort of a warm room or your home via a wireless link, for example. If you would rather not lay out the capital to purchase your own equipment then you can use one of the commercial, remotely operated robotic telescopes. In fact you don't actually have to directly use any such equipment to 'observe' asteroids. There are a number of photographic and image archives available on the Internet that can be searched for these elusive bodies. The professional automated searches have, in the past, recruited members of the public to search their images for new asteroids and given them credit for such discoveries. Although no longer available at present it can be hoped that some of the forthcoming search programs will include such a facility.

As the author did, you can get considerable enjoyment and satisfaction from tracking down and observing asteroids visually with a small telescope or binoculars (and even get an award for your efforts as described in Chap.8). However it must be pointed out at this early stage that, if you wish to progress to making accurate, scientific observations, computer literacy and immediate access to a computer is necessary. The best approach is to have a laptop computer or other device that can be transported to your observing site and on which the necessary data can be displayed. Asteroids, especially fast-moving NEOs, by their very nature, do not stay in the same part of the sky for very long, and their orbits change over the years, so any data you do have will soon be out of date. This is particularly true of newly discovered asteroids whose initial orbits, calculated from a few days' worth of observations, may not be particularly accurate. If you wish to be aware of new discoveries and make further observations of them, then real-time access to the Internet is a must. You can waste an awful lot of paper by printing star charts earlier in the day and then having to throw them away because the sky has clouded over – in England and Maryland at least!

By the time you reach the end of this book you should have a reasonable understanding of the origins, whereabouts, and make-up of asteroids and dwarf planets; the equipment required to observe and image them; the terminology used to describe them; and the knowledge to make astrometric (positional) and photometric (brightness or magnitude) measurements and construct light curves plus monitor occultations of stars by asteroids. Do not discard an observation because it doesn't fit with what has gone before. It may be a fault with your equipment or an error in your analysis, but it could just be real. If you are unsure check with a colleague before going public. There are other books that cover some of these topics in much greater depth, and the Internet is a truly wonderful resource. There is a list of relevant books and websites at the end of this book.

There are numerous local and national astronomical societies, international groups and special Interest mailing lists that welcome newcomers and experienced amateurs alike. Some of these organizations make grants available for the purchase of equipment for specific projects. If you can get to meetings of like-minded amateurs then try to do so, but quite often such gatherings are broadcast live or available in recorded form over the Internet. Sharing your findings and your problems will help you make much faster progress and contribute to your enjoyment of this particular aspect of amateur astronomy. In general, both amateur and professional astronomers respond positively to questions sensibly posed. As your knowledge and experience grow you will then be in a position to return the favor. In Part II you will find examples of amateur activities ranging from simple-to-make visual observations to the most advanced work using CCD and video cameras.

The demise of the amateur astronomer has been rumored for some years now. This is particularly true as far as the subject matter of this book is concerned. The professional automated surveys, especially those yet to become operational, such as the Panoramic Survey Telescope and Rapid Response System (Pan-STARRS) and the Large Synoptic Survey Telescope (LSST), will make us all redundant, won't they? Not so! There are many areas where amateur astronomers can make a real contribution to our knowledge of asteroids. Follow-up observations of new discoveries, light curves, occultations, and measurements of absolute magnitude are as relevant today as they have been for years. Once you have shown yourself to be a competent observer then your results will be as readily accepted as those made by professionals and very much welcomed by those same people. It does no harm to one's sense of well-being to see one's name included in the list of authors at the head of a paper published in a respected refereed journal!

This author's exposure to the asteroid world came sometime in the late 1950s. A popular comic (in the UK) was the 'Eagle,' and one of my favorite characters was an intrepid astronaut (they were called 'spacemen' way back then) by the name of Dan Dare. On one of his travels his spaceship appeared to be on a collision course with an asteroid. Luckily it turned out to be a binary, and they passed safely between the two objects. My second encounter came via a BBC radio series 'Journey into Space,' the lead character being one Jet Morgan. Some of you may

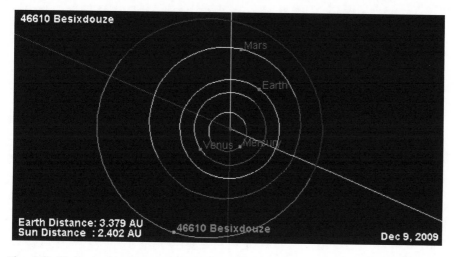

Fig. 1.2. Orbital diagram of asteroid (46610) Besixdouze (Credit: NASA/JPL – Caltech).

believe the NEAR spacecraft was the first to (crash)land on an asteroid, but you would be wrong. Jet Morgan's ship did so, but, judging by the sound effects, his landing was somewhat harder than that of NEAR. Some years later, while working in the United States, I took my two sons to see the film *The Little Prince*, based on the book of the same name, at Radio City Music Hall in New York. The little prince lived on asteroid B-612 from which the B612 Foundation (see Chap.6) takes its name. As this asteroid has active volcanoes it must be a large differentiated asteroid, as described in Chap.4, with a mantle of molten rock – possibly a candidate for dwarf planet status? There is actually a real, Main Belt, asteroid B612 that has the formal designation (46610) Besixdouze (Fig. 1.2) – B612 being the hexadecimal equivalent of that number. Designations and asteroid groups are described in Chaps.2 and 3, respectively.

Before we look at the various ways in which you can observe and image asteroids, let us take some time to examine the nature of asteroids and discover how they came to be where they are in the Solar System. In this first part of the book you will find many references to the role of amateur astronomers as described in Part II.

Chapter 2

Small (and Not So Small) Solar System Bodies

This chapter will cover the somewhat confusing terminology surrounding:

- How planets, dwarf planets, and asteroids are so defined.
- The meaning of the myriad of numbers and names used to designate asteroids.
- The quantities used to describe the orbit of an asteroid around the Sun – its orbital elements.

Planets and Dwarf Planets

It used to be so simple. There were the large objects (the Sun and the planets), the small objects (the asteroids and comets) and the very small (dust, meteoroids, solar wind, cosmic rays, and the like). Pluto, with its eccentric and highly inclined orbit (relative to the other planets), was something of an oddity, but nobody really questioned whether or not it was a planet, at least not until the discovery of a large Edgeworth–Kuiper Belt Object (EKBO) in July 2005. Subsequently numbered and named (136199) Eris, 2003 UB$_{313}$ proved to be slightly larger than Pluto (now numbered 134340). Should this object, informally named 'Xena' at the time of discovery, be considered as the tenth planet? The astronomical world was divided. Some wanted it defined as a planet proper, while others were not so sure. There was much debate as to what such a non-planet should be called, or indeed how planets and asteroids should be categorized.

The matter was resolved at the XXVIth General Assembly of the International Astronomical Union (IAU), which was held in Prague, Czech Republic, during August 2006. Two resolutions, 5 and 6, were passed, but not without considerable discussion relating to planets, asteroids, and comets. The outcome of these resolutions is that the Solar System is now made up of planets, dwarf planets, and small solar system bodies (e.g., asteroids and comets). The formal definitions are:

R. Dymock, *Asteroids and Dwarf Planets and How to Observe Them*,
Astronomers' Observing Guides, DOI 10.1007/978-1-4419-6439-7_2,
© Springer Science+Business Media, LLC 2010

Resolution 5

A planet is a celestial body that:

- Is in orbit around the Sun.
- Has sufficient mass for its self-gravity to overcome rigid body forces so that it assumes a hydrostatic equilibrium (nearly round) shape.
- Has cleared the neighborhood around its orbit.

A dwarf planet is a celestial body that:

- Is in orbit around the Sun.
- Has sufficient mass for its self-gravity to overcome rigid body forces so that it assumes a hydrostatic equilibrium (nearly round) shape.
- Has not cleared the neighborhood around its orbit.
- Is not a satellite.

All other objects, except (natural) satellites, orbiting the Sun shall be referred to as 'small solar system bodies.'

Resolution 6

Pluto is a dwarf planet and is recognized as the prototype of a new category of trans-Neptunian object (TNO). An IAU process will be established to select a name for this category.

In summary a planet is a large round object and a dwarf planet is a small round object. In practice the term 'small solar system bodies' appears to have been still-born. These mostly irregularly shaped bodies are still known, and will probably always be known, as asteroids and comets.

At this time asteroids in orbits similar to that of Pluto were known, informally, as Plutinos. The first of these, 1993 RO, was discovered by Dave Jewitt and Jane Luu in 1993. Such objects make two orbits for every three made by Neptune and are thus said to be in a 3:2 resonance with that planet. Up to 2004 152 of these objects were discovered, and it is estimated that there could be 1,400 with a diameter greater than 100 km.

In June 2008 the IAU introduced the term plutoid – the formal announcement being:

> Plutoids are celestial bodies in orbit around the Sun at a semi-major axis greater than that of Neptune's that have sufficient mass for their self-gravity to overcome rigid body forces so that they assume a hydrostatic equilibrium (near-spherical) shape, and that have not cleared the neighborhood around their orbit. Satellites of plutoids are not plutoids themselves, even if they are massive enough that their shape is dictated by self-gravity.

The three known and named plutoids are (134340) Pluto (136199), Eris, and (136742) Makemake. There are many more large asteroids waiting in the wings to be 'upgraded' to dwarf planet status and, almost certainly, many more orbiting beyond Neptune waiting to be discovered.

You will note that in the IAU announcement concerning plutoids there is no mention of 3:2 resonance with Neptune, merely that the semi-major axis of a plutoid should be greater than that of Neptune. So all plutinos are plutoids, but not all plutoids are plutinos! EKBOs, TNOs, plutinos, and plutoids are discussed in more detail in Chap. 3.

Earlier we used the term 'resolved,' but that is perhaps a little too definitive at the present time! In August 2008 a conference 'The Great Planet Debate: Science as Process' was held at the Johns Hopkins University Applied Physics Laboratory, Maryland. The post-conference press release stated 'Different positions were advocated, ranging from reworking the IAU definition (but yielding the same outcome of eight planets), replacing it with a geophysical-based definition (that would increase the number of planets well beyond eight), and rescinding the definition for planet altogether and focusing on defining subcategories for serving different purposes. No consensus was reached.'

One of the most sensible proposals suggests that in the same way we have various classes of stars, we should have various classes of planets, but that they should all be planets, e.g., Jovian, terrestrial, and dwarf.

Asteroids

For now, we can define planets and dwarf planets, but what of asteroids? *The Encyclopedia of the Solar System*, Second Edition, published by Academic Press in 2007, defines an asteroid as 'A rocky, carbonaceous or metallic body, smaller than a planet and orbiting the Sun.'

Asteroids are by no means all solid bodies, as will be explained in Chap. 4. Those less than 100–150 m in diameter can be considered as solid, while larger ones, between 100 and 300 m or so, are frequently rubble piles, for example (25143) Itokawa visited by the Japanese spacecraft *Hayabusa* in 2005 September and shown in Fig. 2.1. These are the result of the parent bodies being disrupted by impact and then reforming under the influence of gravity – much as planetesimals formed in the early Solar System.

Itokawa shows no outward signs of such an impact, but (2867) Steins certainly does! Figure 2.2 is a series of images obtained by the *Rosetta* spacecraft in September 2008 while on its way to comet 67P/Churyumov-Gerasimenko. The asteroid is approximately 5 km in diameter and was obviously involved in a mighty collision, the crater at the top being of the order of 2 km in diameter. The crater chain

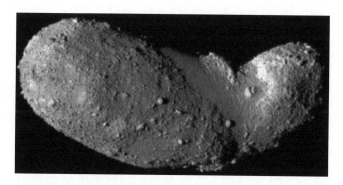

Fig. 2.1. (25143) Itokawa, an example of a rubble-pile asteroid (Credit: JAXA).

Fig. 2.2. (2867) Steins showing multiple impact craters (Credit: ESA 2008 MPS for OSIRIS team).

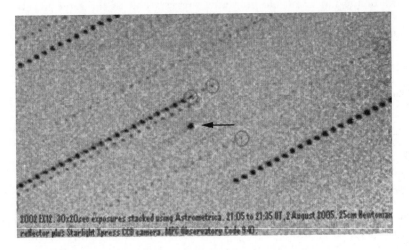

Fig. 2.3. 2002 EX$_{12}$, initially classified as an asteroid but later defined as a comet, 169/P (Credit: *Astrometrica*).

running from top to bottom in the image shows that it suffered further after the main impact. The evidence of the sequence of impacts is that the topmost small crater overlaps the rim of the large one, showing that it occurred after the major impact.

The distinction between asteroids and comets is somewhat fuzzy – pun intended. If an object shows no signs of a coma or tail, then it is usually classed as an asteroid. However some objects initially classed as asteroids have later shown evidence of cometary activity. One such example, shown here imaged in August 2005, is 2005 EX$_{12}$. This was reclassified as a periodic comet, 169/P. Its faint tail can be seen in Fig. 2.3.

This image is actually a number of images stacked to allow for the motion of the object – the stars therefore appearing as lines of dots. The software that makes this possible is *Astrometrica*, which will be discussed in more detail in Chaps.10 and 11.

On the other hand comets, after many orbits around the Sun, eventually outgas all of their volatiles and become extinct. 2003 PG$_3$ may be just such an object. Just to complicate matters further, (5154) Pholus, a Centaur, is most likely a comet nucleus that has never been active.

Designations Old and New

The first asteroids, discovered in the nineteenth century, were given a name without an associated number, e.g., Ceres and Pallas. In 1852 James Ferguson developed a system of numbering asteroids in order of their discovery, namely: ① Astraea, ② Hygiea and ③ Eunomia. There is some debate as to who introduced this system, as Wolf and Gould also claim to have done so in 1851. A few asteroids were also given rather complicated symbols, for example (7) Iris was denoted by the symbol ![symbol], but this method was discontinued because the symbols became hard to draw and recognize. The assignment of a number was made by the editor of the publication *Astromische Nachrichten (AN)*. The problem with this system was that a newly discovered object could be nothing of the sort but merely a further observation of a known asteroid.

In 1892 a system of provisional designations, suggested by Kruger, was implemented with new asteroids identified by their year of discovery followed by a capital letter. The following year the designation was changed to include the year followed by two capital letters. These were to be used consecutively irrespective of any change in year.

The present day system of designations was suggested by Bower in 1924 and implemented in 1925. An example of a provisional designation is 2008 VU$_3$. The first four numbers are the year of discovery, the next letter indicates the half-month period during which the object was discovered, and the final letter and number the order of discovery within that period. The periods are shown in Table 2.1.

For example, the first 25 asteroids discovered during the half month period November 1–15 in 2008 will be numbered 2008 VA to 2008 VZ. The next 25 will be numbered 2008 VA$_1$ to 2008 VZ$_1$. For subsequent discoveries in that period the subscript number will be 2, 3, 4, and etc. There have been other designations, used by special surveys, for example, and during wartime when the discoverers were unable to communicate their findings to the appropriate body.

Observations reported to the Minor Planet Center (MPC) include the provisional designation in packed format. So 2008 TT$_{26}$ becomes K08T26T, the first two digits of the year being indicated by the letter K. Reporting will be explained in greater detail in Chap. 11.

Table 2.1. Half month of discovery

Letter	Half month	Letter	Half month
A	Jan 1–15	N	Jul 1–15
B	Jan 16–31	O	Jul 16–31
C	Feb 1–15	P	Aug 1–15
D	Feb 16–19	Q	Aug 16–31
E	Mar 1–15	R	Sep 1–15
F	Mar 16–31	S	Sep 16–30
G	Apr 1–15	T	Oct 1–15
H	Apr 16–30	U	Oct 16–31
J	May 1–15	V	Nov 1–15
K	May 16–31	W	Nov 16–30
L	Jun 1–15	X	Dec 1–15
M	Jun 16–30	Y	Dec 16–31

Numbering and Naming

When an asteroid is first discovered it may be assigned a temporary designation by the discoverer. Upon confirmation of that discovery, by a further night's observations for example, it will be assigned a provisional designation by the MPC. Discovery confirmation is an area in which amateur astronomers can successfully partake, as will be described in Chap. 11. The MPC, set up in 1947 and operating under the auspices of the IAU, is responsible for collecting observational data (astrometry and photometry) for asteroids and comets and calculating their orbits.

A permanent number can be assigned once the orbit of an asteroid is well-defined. A newly discovered Main Belt asteroid must be observed on two or more nights at four oppositions, or orbits, but an NEO may be assigned a number after being observed during only two or three oppositions. For example 2001 VZ_{87}, discovered by the Near Earth Asteroid Tracking observatory (NEAT) in November 2001 was subsequently numbered 111118.

The discoverer can propose a name after the object receives a permanent number, and such proposals are vetted by the Committee on Small Body Nomenclature (CSBN) of the IAU. To add to the confusion discoverers have, on occasions, given an unofficial name to their newfound object. For example the satellite of 2003 UB_{313} (unofficially named Xena) was given the name Gabrielle by its discoverer Mike Brown, and, subsequently, this object metamorphosed into (136199) Eris I (Dysnomia). As demonstrated by that designation, satellites of asteroids and dwarf planets are given a number – I, II, III, etc. – after the name of their parent body as well as a name.

A fuller description of the numbering and naming sequence of events can be found on the IAU website page 'Naming Astronomical Objects' and on the MPC website page 'Guide to Minor Body Astrometry.'

Asteroid Orbits

The word 'orbit' has already been mentioned several times in this book, so it is perhaps time to elaborate on this subject and why up-to-date knowledge of the orbits of asteroids is important, and in the case of near-Earth asteroids, essential.

The path followed by an asteroid around the Sun (or any solar system body circling another) is defined by six numbers (the orbital elements) plus the epoch (date) for which those numbers are valid, as described in Table 2.2 and shown in Fig. 2.4. The column headed 'MPC notation' lists the abbreviations used by that organization. An example for asteroid (35396) 1997 XF_{11} follows.

The closest point of the orbit to the Sun, perihelion, can be calculated using the formula:

$$q = a(1 - e) \text{ and the furthest point, aphelion, is given by } Q = a(1 + e).$$

Orbital elements and orbit diagrams are freely available from the websites of the MPC and Jet Propulsion Laboratory (JPL). The data following was obtained from the MPC's Minor Planet Ephemeris Service website (for the author's observatory, code 940, the significance of which will be explained in Chap. 11). The meaning of

Table 2.2. Orbital elements

Name	Symbol	MPC notation	Description
Mean anomaly	M	M	Although the true definition is a little more complicated, this is essentially the current angular distance from perihelion to the present position of the asteroid measured in the direction of motion
Semi-major axis	a	a	(Half) the length of the long axis of the ellipse
Eccentricity	e	e	A measure of the deviation of the orbit from a circle (all asteroid orbits are ellipses) $e = c/a$. For a circle $e = 0$, and for a typical Main Belt asteroid $e = 0.1–0.2$
Inclination	i	Incl	The angle between the plane of the orbit of the asteroid and the ecliptic. If the inclination is $>90°$ then the motion of the object is considered to be retrograde
Longitude of the ascending node	Ω	Node	The direction in space of the line where the orbital plane intersects the plane of the ecliptic. It is measured eastwards (increasing RA) from the vernal equinox (first point of Aries)
Argument of perihelion	ω	Peri	Defines how the major axis of the orbit is oriented in the orbital plane and is the angle between the ascending node and the perihelion point measured in the direction of motion
Epoch		Epoch	The date on which a set of orbital elements were calculated

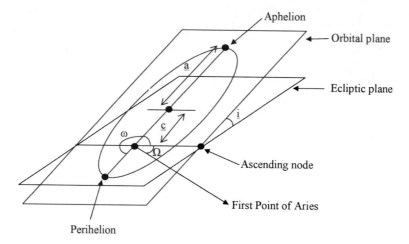

Fig. 2.4. Orbital elements (Art by the author).

the magnitude parameters, H and G, will be covered in Chap. 14. The Julian Date system (JDT below) was introduced in 1582 by the French scholar Joseph Justus Scaliger, who named it in honor of his father, Julius Caesar Scaliger. The starting date is January 1, 4713 b.c., which predates all known astronomical records. Time, after the decimal point, is measured in fractions of a day starting at noon – midnight being 0.5 of a day.

(35396) 1997 XF11								
Epoch 2002 Nov. 22.0 TT = JDT 2452600.5						MPC		
M	41.47384	(2000.0)			P		Q	
n	0.56889082	Peri.	102.64552		+0.72558616		+0.68697184	
a	1.4425031	Node	213.98824		−0.65582495		+0.67278524	
e	0.4841203	Incl.	4.09603		−0.20837072		+0.27464471	
P	1.73	H	16.9	G	0.15	U	2	

Table 2.3. Relationship of data and diagrams

Orbital element	MPC data	Notation
Mean anomaly (M)	M	Angle PSM
Semi-major axis (a)	a	Half the distance PSA
Eccentricity (e)	e	((AP/2)-SP)/(AP/2)) or c/a in Fig. 2.4
Longitude of the ascending node (Ω)	Node	Angle FSN
Argument of perihelion (ω)	Peri	Angle NSP

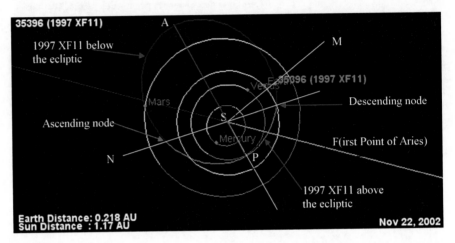

Fig. 2.5. Orbit diagram of asteroid 35396, 1997 XF11 (Courtesy NASA/JPL-Caltech, with additional data and graphics by the author)

Figure 2.5 shows the orbit of asteroid 35396, previously known by its provisional designation, 1997 XF_{11}.

Table 2.3 relates the orbital elements and MPC data listed in Table 2.2 to the notation in Fig. 2.5.

In addition to the orbital elements described here an ephemeris, a list of predicted positions of a celestial object, can also be obtained from the MPC and other websites listed in Appendix B of this book. Software packages such as *Find_Orb*, *Exorb*, or *CODES*, enable observers to generate their own orbital elements and ephemerides from a set of observations formatted to MPC requirements (an explanation of which can be found in Chap. 11). The projected track of an asteroid can be generated using software such as *Guide* (again from Project Pluto) or *Megastar* from Willmann-Bell. *SOLEX*, written by Aldo Vitagliano, is a free software package that can model many, many aspects of the motions of asteroids and dwarf planets. If you would like to know how Carl Gauss determined the orbit of (1) Ceres at the beginning of the nineteenth century, and possibly try it for yourself, then check the website listed in Appendix B of this book.

An asteroid is always subject to the gravitational forces exerted by the planets and even other asteroids. These perturbations, as they are known, cause its orbit to change gradually over time. The orbital elements used in this example, and generally quoted, are known as osculating elements, as they describe the path the body would follow at the given epoch if the perturbations were to cease at that time. Similarly the orbit derived from these elements is known as the osculating orbit.

The term 'proper elements' refers to a set of orbital elements calculated by ignoring perturbations and can simply be described as representing the average motion of the body concerned. These were developed by Hirayama in 1918 and have proved extremely useful in classifying asteroid families, which are further described in Chap.3.

Lost? Perhaps Not

If an asteroid has not been observed for as little as 2 years its actual position may differ from its predicted position by several arc minutes. The 'Follow-Up Astrometric Program,' run by the Italian Organization of Minor Planet Observers, has the objective of observing asteroids that are in danger of becoming lost. Similarly the Lowell Observatory's 'Hierarchical Observing Protocol for Asteroids' includes a selection criterion, 'Danger of loss.' These facilities can be used to select asteroids for observation and are described further in Chap.11

So you now have some idea as to what an asteroid is and what a dwarf planet is, how they are designated, and the way in which their orbits are described. The latter in particular should be useful when reading the next chapter, describing where the various groups of asteroids, and dwarf planets, are found in the Solar System.

Chapter 3

Groups and Families

This chapter describes the present locations of asteroids in our Solar System and mentions recent discoveries in other planetary systems. How the asteroids and dwarf planets arrived in these various locations and their continuing evolution is described in Chap. 5.

Asteroid Groups

The various groups of asteroids in the Solar System are (in order of increasing distance from the Sun):

- Vulcanoids
- Near-Earth asteroids/objects (NEA/Os)
- Main Belt asteroids
- Trojans
- Centaurs
- Edgeworth–Kuiper Belt objects (EKBOs) /trans-Neptunian objects (TNOs)

Some of these groups are further subdivided, as will be described in this chapter. The number of asteroids discovered increases by several thousand per month, but it should be noted that as we move farther from the Sun the detection of smaller and/or darker objects becomes more difficult if not impossible, so there may be many which remain undiscovered.

Those interested in the history of discovery and the reasons behind the sometimes rather unusual names might like to refer to the *Dictionary of Minor Planet Names* by Lutz D. Schmadel, published by Springer. Asteroids are often named for someone who has performed sterling work in the field of astronomy, for example asteroid (6137) Johnfletcher is named after a UK amateur astronomer.

Table 3.1 lists the orbital elements and other data for a selection of asteroids from some of the groups described in this chapter. Orbital elements, explained in Chap. 2, are always quoted for a given date or epoch. It is advisable to obtain the latest and most accurate orbital elements, specific to your location and time of observation, from the Minor Planet Center when attempting to locate an asteroid or dwarf planet.

R. Dymock, *Asteroids and Dwarf Planets and How to Observe Them*,
Astronomers' Observing Guides, DOI 10.1007/978-1-4419-6439-7_3,
© Springer Science+Business Media, LLC 2010

Table 3.1. Orbital elements and other data for a selection of asteroids

	(1862) Apollo	(253) Mathilde	(624) Hektor	(2060) Chiron	(15874)
Group	NEO	Main Belt	Jupiter Trojan	Centaur	TNO (scattered disk object)
Date of discovery	April 24, 1932	November 12, 1885	February 10, 1907	October 18, 1977	October 9, 1996
Epoch	November 30, 2008	November 30, 2008	November 30, 2008	November 30, 2008	November 30, 2008
Eccentricity	0.56	0.27	0.02	0.38	0.58
Semi-major axis (AU[a])	1.47	2.65	5.23	13.71	83.33
Perihelion distance (AU)	0.65	1.94	5.12	8.51	184.34
Inclination (°)	6.35	6.74	18.18	6.93	23.99
Longitude of ascending node	35.75	179.62	342.80	209.29	217.77
Argument of perihelion	285.83	157.51	183.83	340.02	184.34
Mean anomaly	286.38	249.31	218.27	90.75	3.59
Period (years)	1.78	4.3	11.97	50.76	760.72
Aphelion distance (AU)	2.29	3.35	5.35	18.91	131.65

[a]AU astronomical unit; originally the average distance of Earth from the Sun, 149,597,870 km (92,975,681 miles). A more recent, and more complicated, method of calculation defines Earth's average distance from the Sun as 1.000000031 AU

Vulcanoids

Vulcan is the name given to a hypothetical planetary body once believed to have orbited closer to the Sun than Mercury. After searching for the past 150 years astronomers have yet to find any such object. It is believed that such an object could exist if it were in a very circular orbit between 0.07 and 0.21 AU from the Sun. In 1998 a search was conducted by G. Schumacher and J. Gay using the LASCO coronograph on board the SOHO spacecraft. The conclusion was that no body with a diameter greater than 60 km existed within the area investigated.

Although no Vulcanoids have been detected using the cameras on SOHO, many comets have been discovered in this way by both professional and amateur astronomers.

NASA's *MESSENGER* spacecraft, which is presently in orbit around the Sun and will be the first to orbit the planet Mercury, is also being used to search for Vulcanoids. Whereas the *SOHO/LASCO* project could detect any object of 60 km in diameter or larger, the *MESSENGER* imaging team believe they can detect objects as small as 15 km.

Near-Earth Asteroids/Objects (NEAs/Os)

Although strictly near-Earth asteroids (NEAs), these are nearly always referred to as near-Earth objects (NEOs), and with the odd exception, this latter term will be used throughout this book. NEOs are divided into three groups defined by their orbits with respect to that of Earth and named after the first in each group to be discovered. As shown in Fig. 3.1, the Atens spend most of their time within Earth's orbit, the Apollos spend most of their time further from the Sun than Earth, and the Amors always stay outside Earth's orbit.

The first asteroid known to have an Earth-crossing orbit, (1862) Apollo, was discovered by Karl Reinmuth on April 24, 1932. (1221) Amor was discovered on March 12 by E. Delporte in that same year, and (2062) Aten was discovered on

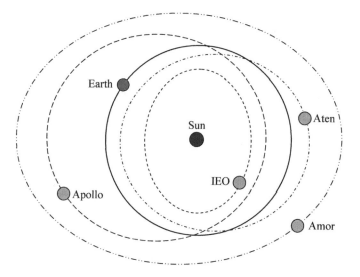

Fig. 3.1. Orbits of NEOs (Art by the author).

Fig. 3.2. Inner Earth Object 2008 UL$_{90}$ (Credit: Peter Birtwhistle, Great Shefford Observatory, UK).

January 7, 1976, by Eleanor Helin. The total number of NEOs so far discovered is approaching, and by the time you read this will most likely have passed, 6,000.

There is a subset of the Atens known as inner-Earth objects (IEOs), and informally known as the Apoheles. All of the orbit of an IEO lies within that of Earth, and this makes them extremely difficult to detect. Not only must an IEO have a semi-major axis smaller than that of Earth, but both its aphelion (Q) and perihelion (q) distances must be less than Earth's perihelion distance (0.9833 AU). There are 14 known or suspected asteroids in this class, the first to be discovered being 1998 DK$_{36}$ and the most recent 2008 UL$_{90}$. Figure 3.2 shows an image of the latter, circled, obtained by UK amateur astronomer Peter Birtwhistle. The bright, partially obscured star in the image is TYC 829-1743-1.

The term Arjunas is sometimes used for asteroids with Earth-like orbits, e.g., low eccentricity, low inclination, and a semi-major axis close to 1 AU. To some extent this class overlaps and includes asteroids in the three previously mentioned classes of NEOs – Atens, Apollos, and Amors. The first to be discovered, by the *Spacewatch* automated search program, was 1991 VG on November 6, 1991.

The term 'Mars crosser' applies to those asteroids that cross the orbit of Mars but not the orbits of any of the other terrestrial planets. The first Mars crosser to be so identified was 132 Aethra, discovered by James Craig Watson in 1873.

The Main Belt

The Main Belt of asteroids lies between Mars and Jupiter and stretches from approximately 2.1 to 3.3 AU from the Sun. Zone I (Inner) stretches from 2.1 to 2.5 AU, Zone II (Central) from 2.5 to 2.8 AU, and Zone III (Outer) from 2.8 to 3.3 AU. The first asteroid to be discovered, Ceres, was found by Giuseppe Piazzi on January 1, 1801. Further discoveries followed: Pallas in 1802, Juno in 1804, Vesta in 1807, and a total of fifteen (all Main Belt asteroids) by the end of 1851.

Figure 3.3 shows a selection of typical Main Belt asteroids to the same scale, imaged by passing spacecraft. The visible part of Mathilde is 59 km wide by 47 km. The manner in which craters on our Moon and other planets are named is common knowledge, but craters on asteroids have also been named by the International Astronomical Union (the only organization that can officially name celestial objects). The large crater in the center of Mathilde is named Karoo, after a coal basin on Earth. The three large and many small craters on this asteroid indicate a history of heavy bombardment, which led to the breakup of some asteroids and creation of the resulting families. Such families, e.g., Flora, Eos, Koronis, and Themis are defined by their orbital characteristics. They are often referred to as the Hirayama families, after Kiyotsugu Hirayama who discovered these orbital similarities in 1918. How these families came to be will be covered in Chap. 5.

The popular press often depicts the Main Belt as a place crowded with asteroids virtually jostling one another for space. It is more likely that, on average, such objects

Mathilde   Gaspra   Ida

Fig. 3.3. Three Main Belt asteroids (Credit: NASA/NSSDC/NEAR/Galileo).

are of the order of 500,000 to 1,000,000 km apart. So, rather than being in a rush-hour traffic jam, you would feel very much alone on this particular astral highway.

Within the Main Belt there are gaps, corresponding to various orbital period ratios, or resonances, with Jupiter, e.g., 4:1, 3:1, 5:2, and 2:1. Such gaps are not necessarily completely devoid of asteroids but are sparsely populated. For example in the 2:1 (Hecuba) gap at 3.27 AU there are three groups of asteroids known as Zhongguos and Griquas, both of which are in relatively stable orbits, and an unnamed group whose orbits are unstable. The gaps, and what happens to asteroids that drift in to them, are explained in greater depth in Chap. 5.

Not all Main Belt residents are of the asteroid variety. In 2006 David Jewitt and Henry Hsieh reported that asteroid 118401 had a comet-like dust tail. This was the third Main Belt asteroid, in a near circular orbit, to exhibit such activity. It was previously thought that asteroids in this region were too close to the Sun to retain ice. It is possible that a thick crust could protect an icy interior with the occasional impact making a hole in the crust and allowing the ice to sublimate. Alternatively, but less likely, the bodies could have originated further out and been deflected into their present orbits.

Observing Main Belt asteroids might seem a bit 'ordinary' and less of a challenge compared with the excitement generated by a newly discovered NEO whizzing by Earth or the search for large bodies in the far reaches of the Solar System. However it is a good place for beginners to start, as described in Chaps. 8 and 9, as these asteroids are relatively bright and their orbits, and therefore positions, well defined. There are also opportunities for the more advanced amateur astronomer in this area. For example only a small proportion of such asteroids have well-defined lightcurves and accurate values of absolute magnitude. The photometry involved in defining these values will be explored in Chaps. 13 and 14.

Trojans

Locations and Numbers

Trojan asteroids occupy the same orbit as their parent planet but are located around the Lagrangian L4 and L5 points, 60° ahead and 60° behind the planet, as shown in Fig. 3.4.

The numbers of Trojans associated with the various planets are shown in Table 3.2. No Trojans have so far (November 2008 Minor Planet Center data) been discovered around other planets in the Solar System.

Saturn and Uranus have no Trojans because the gravitational forces associated with the presence of massive planets on both sides of them prevent asteroids from congregating at the L4 and L5 points for any length of time. It may be possible for Earth to have Trojans, but, as yet, none has been discovered.

Martian Trojans

The first Martian Trojan, (5261) Eureka, was discovered by H. E. Holt and D. H. Levy on June 20, 1990. It shares Mars's orbit and is located at the Lagrangian L5 point.

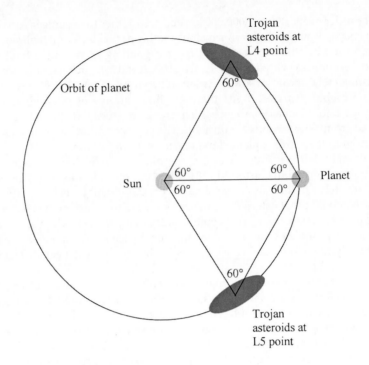

Fig. 3.4. Location of Trojan asteroids (Diagram by the author).

Table 3.2. Numbers of Trojan asteroids

Parent planet	L4 Trojans	L5 Trojans
Mars	1	3
Jupiter	1618	1274
Neptune	6	0

Jupiter (or Jovian) Trojans

The vast majority of Trojan asteroids, all named after Trojan war heroes, are those associated with Jupiter. The first, (588) Achilles, was discovered by Max Wolf on February 22, 1906. Jupiter Trojans oscillate, or librate, around the Lagrangian L4 and L5 points by, typically, ±15–20°. Estimates suggest that there may be as many as 600,000 Jupiter Trojans larger than 1 km in diameter, which is approximately equal to the projected number of Main Belt asteroids of the same size. The largest (Jupiter) Trojan is (624) Hektor, which measures 370 × 195 km.

Neptune Trojans

The first Neptune Trojan, 2001 QR$_{322}$, was discovered on August 21, 2001, in the course of the Deep Ecliptic Survey, a NASA-funded survey of the outer Solar System.

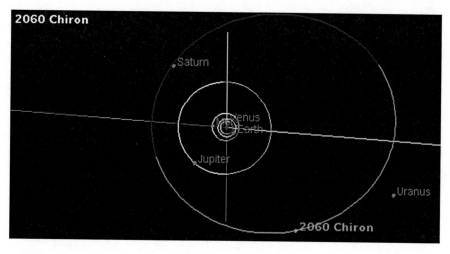

Fig. 3.5. (2060) Chiron – the first Centaur to be discovered (Credit: NASA/JPL-Caltech).

Centaurs

Centaurs occupy the region of the Solar System between Jupiter and Neptune, having a semi-major axis between 5.5 and 29 AU. At times they behave like comets, with them being surrounded by a small coma of gas, and at others they behave like regular asteroids, completely inert. Due to their large size (hundreds of kilometers in diameter) and location they could pose a threat to life on Earth. They are subject to the gravitational influences of the giant planets and could have their orbits changed to such an extent that they are flung in to the inner Solar System (or, if we are lucky, out of).

The first Centaur, 1977 UB, subsequently (2060) Chiron (Fig. 3.5), was discovered by Charles Kowal in October 1977 using the 1.15-m Schmidt telescope on Mt. Palomar. Its orbit takes it close to both Saturn and Uranus. The largest Centaur discovered so far is (10199) Charliko, which is 260 km in diameter. As of December 2008, 242 Centaurs and scattered disk objects (SDKs) were known.

The Edgeworth–Kuiper Belt

This group of asteroids, 1,093 discovered by the end of 2008 and also known as trans-Neptunian objects (TNOs), can be subdivided into:

- Plutinos (an informal name)
- Plutoids
- Classical EKBOs
- Scattered disk objects (SDOs)
- Detached objects (previously known as extended SDOs)

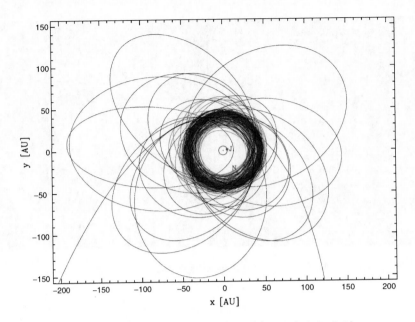

Fig. 3.6. The Edgeworth–Kuiper Belt. *Red* = Plutinos, *Blue* = Classical EKBOs, *Black* = SDOs (Credit: Dave Jewitt).

Figure 3.6 illustrates some of these various subdivisions.

The exploration of the Solar System beyond Neptune started with the discovery of Pluto by Clyde Tombaugh on February 18, 1930. The subsequent, and still controversial, reclassification of Pluto as a dwarf planet is described in Chap. 2. In the same year that Pluto was discovered Frederick Leonard thought there might be debris beyond the orbit of Neptune left over from the formation of the planets. He coined the term 'planetesimals' for such objects. In 1943 Irish amateur astronomer Kenneth Essex Edgeworth suspected that the disk of planet-forming material continued beyond the planetary orbits but that its density would be very low. This might allow small bodies to form, but they would be too few and far between to coalesce into planets. In 1951 Gerald Kuiper suggested many small bodies might have formed beyond Neptune as the planets came into being. The relative sizes of Pluto and other recently discovered, large EKBOs/TNOs, some of which have been classified as dwarf planets, are shown in Fig. 3.7.

That the EKB is a fairly empty place is supported by two recent studies. The first showed that the orbit of Halley's Comet was unaffected by the gravitational influence of EKBOs. The second studied the possible effect of thermal emission from the EKB on measurements of cosmic microwave background radiation – no effect being observed. There may nevertheless be something strange lurking in the outer Solar System. The trajectories of the spacecraft *Pioneer 10* and *Pioneer 11*, launched in 1972 and 1973, respectively, are not quite as they should be. That is, they are slowing down more than predicted. Many ideas have been put forward as to the cause, including the presence of as yet undetected EKBOs.

Plutinos and Plutoids

As the (unofficial) name suggests, plutinos have similar orbits to Pluto in that they have a semi-major axis between 39 and 40 AU, a perihelion distance less than or equal to that of Neptune, and are in a 3:2 resonance with that planet. 1993 RO and 1993 RP

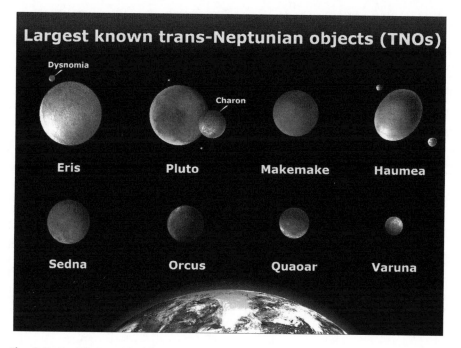

Fig. 3.7. Large EKBOs/TNOs and dwarf planets (Credit: NASA/STScI).

were the first objects in this category to be found – discovered by Dave Jewitt and Jane Luu in September 1993.

As mentioned in Chap. 2, plutoids are dwarf planets with a semi-major axis greater than that of Neptune. The three known and named plutoids are Eris, Makemake, and Pluto.

Classical Edgeworth–Kuiper Belt Objects

Most classical EKBOs move in low inclination; have low eccentricity orbits, mostly between 42 and 45 AU from the Sun; have a semi-major axis between 42 and 48 AU; and occupy various resonances with Neptune, e.g., 5:3, 7:4, and 2:1. Orbital dynamicists describe objects with these characteristics as 'cold.' The first of these to be discovered was 1992 QB$_1$ (giving rise to their informal name of Cubiwanos). It was found by Dave Jewitt and Jane Luu in September 1992. Previous discoveries, such as Pluto itself, were made using photographic plates and blink comparators. These were some of the earliest asteroid discoveries using the then-new technologies of CCDs with computers to do the blinking. The further discovery of large objects in this region put Pluto in its proper context and eventually led to its reclassification as a dwarf planet.

A few classical EKBOs move in highly inclined orbits but are still considered members of this class, as their orbits are roughly circular. These are referred to as 'hot.' As is usual when you try to classify asteroids, there are always some that won't be shoehorned into any particular category. 2008 KV$_{42}$ fits this bill, as its extremely high inclination of 104° might suggest it is a 'hot' EKBO, but its highly elliptical orbit with an eccentricity of 0.56 puts it outside that particular realm.

Scattered Disk Objects

SDOs are asteroids in highly eccentric orbits with semi-major axis greater than 48 AU. Their highly elliptical orbits take them hundreds of AU from the Sun at aphelion. The first SDO, 15874, initially known as 1996 TL_{66}, was discovered by Dave Jewitt, Jane Luu, and Chad Trujillo in October 1996.

Detached Objects

Previously known as extended scattered disk objects, detached objects, of which (90377) Sedna, originally known as 2003 VB_{12}, was the first to be discovered, exist at great distances from the Sun, and move in highly eccentric orbits. For example Sedna has a semi-major axis of 495 AU, an aphelion distance of 914 AU, and an eccentricity of 0.85. The 'detached' in the name indicates that they are both remote from Neptune's gravitational influence and the rest of the EKB. They are so far from other EKBOs that they are sometimes referred to as inner Oort Cloud objects.

And Finally...

News stories, in the UK at least, often finish their bulletins with the above words. In the case of the Edgeworth–Kuiper Belt it will probably be some time before astronomers can utter the same. Even large objects in this region are extremely faint and relatively slow moving, making their detection very difficult.

Detection of smaller objects may be possible by indirect means. Recent observations by Australian, French, and Taiwanese astronomers suggest that many icy bodies as small as 10 m in diameter may exist in the EKB. The not necessarily conclusive evidence for this comes from observations of stars that show random, short-lived dips in brightness at both visible light and X-ray wavelengths. It is believed such dips are due to occultations of the stars by the small EKBOs. (How amateurs may observe occultations of stars, usually by Main Belt asteroids, is described in Chap. 15.)

Astronomers have long searched for a Planet X – a body always just beyond the detection capabilities of their telescopes and imagers. As a former director of the Asteroids and Remote Planets Section of the British Astronomical Association, this author occasionally received e-mails querying the existence of a such a body in the outer Solar System beyond Neptune. A paper by Patryk S. Lykawka and Tadashi Mukai published in the *Astronomical Journal* proposed that "the orbital history of an outer planet with tenths of Earth's mass can explain the trans-Neptunian belt orbital structure." We still await the discovery of such a body, so watch for this, or rather that, space! If you think that such work is beyond the scope of the amateur, then read about the survey, in Chap. 11, being conducted by Eamonn Ansbro.

Moons

Classical Orbits

This "invented" description is useful for distinguishing such satellites from those with complex orbits as described in the next section. The two moons of Mars, Phobos and Deimos, and eight of Jupiter's outer satellites may be captured asteroids. How they might have been captured is described in Chap. 5.

Planetary satellites would be expected to be spherical in shape and orbit their parent bodies with prograde motion (i.e., in the same direction as the planet rotates). The plane of the orbit should also be aligned with the planet's equator. Any small moon that does not obey these 'rules' is likely to be a captured asteroid. Several of the moons of the giant planets exhibit these characteristics, e.g., Sinope (satellite of Jupiter) and Phoebe (Saturn). The large satellite of Neptune, Triton, also exhibits unusual (retrograde) motion, which could be explained if it had been captured by that planet from a solar orbit being, possibly, one half of a binary system – its partner having been flung away from Neptune in the process.

Quasi-Satellites and Horseshoe Orbits

Quasi-satellites are so called because they spend part of their time in orbit around Earth and part in orbit around the Sun. Asteroid (3753) Cruithne is one of three known asteroids of this type, and 2002 AA$_{29}$ follows a horseshoe-shaped orbit around the Sun at an almost identical distance to that of Earth. A simplified version of such an orbit is shown in Fig. 3.8. At point A the asteroid is traveling

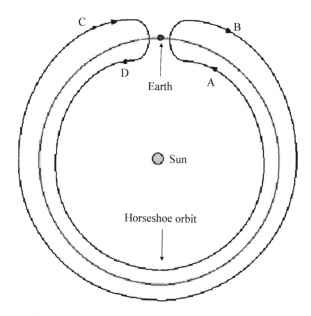

Fig. 3.8. Horseshoe orbit (Diagram by the author).

slightly faster than Earth and, as it approaches, it is pulled into a larger orbit at B and falls behind Earth until it reaches point C. Here it is switched into a smaller orbit by Earth's gravitational attraction and speeds up until it again approaches Earth at point A.

Dwarf Planets

Table 3.3 lists the orbital elements and other data for the known dwarf planets. Orbital elements, explained in Chap. 2, are always quoted for a given date or epoch. As stated above, it is advisable to obtain the latest and most accurate orbital elements, specific to your location and time of observation, from the Minor Planet Center when attempting to locate an asteroid or a dwarf planet. These, and therefore the position of the object, do change over time. The dwarf planets are described in more detail in Chap. 4.

Table 3.3. Orbital elements and other data for the currently known dwarf planets

	(1) Ceres	(134340) Pluto	(136108) Haumea	(136199) Eris	(136472) Makemake
Provisional designation	–	–	2003 EL$_{61}$	2003 UB$_{313}$	2005 FY$_9$
Date of discovery	January 1, 1801	January 23, 1930	March 7, 2003	October 21, 2003	March 31, 2005
Epoch	November 30, 2008	September 22, 2006	November 30, 2008	November 30, 2008	November 30, 2008
Eccentricity	0.08	0.25	0.20	0.44	0.16
Semi-major axis (AU)	2.77	39.45	43.13	67.90	45.43
Perihelion distance (AU)	2.55	29.57	34.72	38.29	38.10
Inclination (°)	10.59	17.09	28.22	44.02	29.00
Longitude of ascending node	80.40	110.38	122.10	35.96	79.57
Argument of perihelion	72.90	112.60	239.18	151.52	295.15
Mean anomaly	344.55	25.25	202.68	198.86	151.60
Period (years)	4.60	247.74	283.28	559.55	306.17
Aphelion distance (AU)	2.99	49.32	51.54	97.52	52.75

Exosolar Asteroids

Dust disks and possible asteroid belts have been detected around other stars. Spitzer Space Telescope observations of Epsilon Eridani have led to the discovery of two asteroid belts and one comet belt around this star. The arrangement of belts and known and possible planets is shown in Fig. 3.9. It is possible that as yet undetected planets are responsible for the position and shape of these belts.

Summary

How many asteroids are there? Current estimates put the total number of asteroids in the Solar System with a diameter greater than 1 km at between 1.1 and 1.9 million. By November 2008 the total number of asteroids discovered had reached a

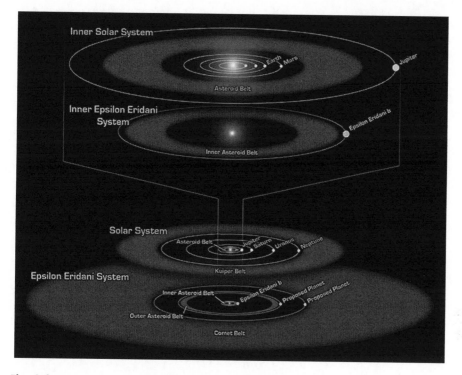

Fig. 3.9. Asteroid and comet belts around Epsilon Eridani (Credit: NASA/JPL-Caltech).

total of 434,524. Of these 200,083 had orbits defined well enough to be numbered. The vast majority of known asteroids, over 90%, reside in the Main Belt. As an indication of the rate at which new asteroids are being discovered, the equivalent totals in November 1998 were 45,399 and 9,709, respectively (Minor Planet Center data).

The reasons for this rapid increase are several: the move from early visual observations to the use of photographic plates and then CCD imaging; the implementation of professional automated surveys using moving object detection software; the use of the Internet to enable immediate notification of potential discoveries needing follow-up observations; and the increasing ability of amateur astronomers to detect fainter and fainter objects. The latter will be discussed further in Chaps. 10 and 11.

Having described the various groups of asteroids it is now time to go into a little more detail as to their actual structure, as described in the next chapter.

Chapter 4

The Nature of Asteroids and Dwarf Planets

This chapter looks at the structure of asteroids and dwarf planets: solid bodies and rubble piles, binary and triple asteroids, how their birthplace affects their make-up, surface morphology, and interior structure. Our knowledge of the size and composition of asteroids is built up from a number of sources:

- Spectroscopy
- The study of meteorites
- Radar observations
- Spacecraft missions to asteroids
- Occultations of stars by asteroids
- Lightcurve photometry

Asteroids vary in both size and shape. 'Diameters' range from meters to hundreds of kilometers (at around 1,000 km we are into the realm of dwarf planets). The word 'diameters' is in quotes because asteroids come in all shapes, from the roughly circular to something your dog might be pleased to receive (Fig. 4.1). (216) Kleopatra measures $217 \times 94 \times 81$ km and may be one object or a very close binary. Radar observations suggest this asteroid may be part of the metallic interior of a larger differentiated parent body.

The physical characteristics of a selection of asteroids from various groups are listed in Table 4.1.

Rubble Piles or Solid Bodies?

Asteroids less than 100–150 m or so in diameter are almost certainly solid bodies (monoliths) – most likely fragments of larger asteroids broken up by collisions. Objects larger than this and under around 300 m in diameter are loosely bound collections of fragments usually described as 'rubble piles.' Their rotation periods are approximately equal to or less than 2.5 h – any faster and they would disintegrate. Like the smaller bodies these would have formed as the result of a collision, but, in such cases, the fragments re-formed into a single object. Figure 4.2 shows a possible scenario for the formation of asteroid (25143) Itokawa. In this case the debris from the collision re-assembled into what is known as a contact binary.

R. Dymock, *Asteroids and Dwarf Planets and How to Observe Them*, Astronomers' Observing Guides, DOI 10.1007/978-1-4419-6439-7_4, © Springer Science+Business Media, LLC 2010

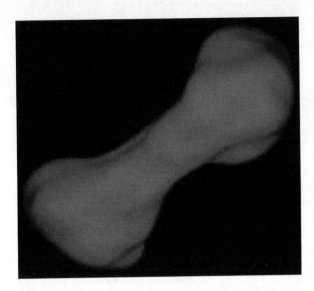

Fig. 4.1. Asteroid (216) Kleopatra (model derived from radar images) (Credit: NASA/JPL).

Table 4.1. Physical characteristics of a selection of asteroids

	(1862) Apollo	(253) Mathilde	(624) Hektor	(2060) Chiron	(15874) 1996 TL$_{66}$
Group	NEO	Main belt	Jupiter Trojan	Centaur	TNO (scattered disk object)
Date of discovery	April 24, 1932	November 12, 1885	February 10, 1907	October 18, 1977	October 9, 1996
Dimensions (km)	1.7 (diameter)	66 × 48 × 46	370 × 195	233 ± 14 × 132–152	~575 ± 115(diameter)
Mass (kg)	5.1×10^{12} (?)	1.03×10^{17}	~1.4×10^{19}	~1×10^{19} (?)	~2×10^{20} (?)
Density (g/cc)	2.0 (?)	1.3	2.0 (?)	?	2.0 (?)
Albedo	0.21	0.04	0.03	0.08 ± 0.1	0.04
Mean surface temp (K)	~222	~174	~122	~75	~31
Rotational period	3.07 h	17.40 days	6.92 h	5.92 h	?

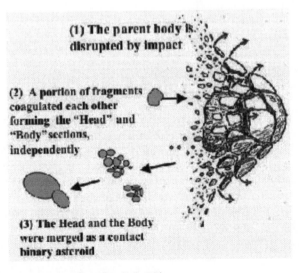

Fig. 4.2. A rubble pile asteroid reforming after a collision (Credit: JAXA).

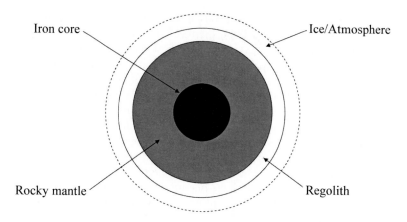

Fig. 4.3. A differentiated asteroid (Diagram by the author).

Amateur astronomers have been able to measure the rotation rates of some rapidly rotating asteroids, thus determining that they are solid bodies rather than rubble piles that would have flown apart. Chapter 13 describes one such project by UK amateur astronomer and director of the Asteroids & Remote Planets Section of the British Astronomical Association, Dr. Richard Miles.

The larger asteroids and dwarf planets – bodies of more than a few hundred kilometers in diameter – will be more or less spherical and, most likely, differentiated, consisting of layers of material as shown in Fig. 4.3. The spherical shape is a result of gravitational forces, and the differentiation is due to heating, by the decay of radioactive elements, during formation – the heavier elements sinking to the center of the body due to the force of gravity.

Not all large asteroids and dwarf planets will include all of these features – for example only Pluto is known to have a significant atmosphere. The regolith is a layer of dust and fragments of rock caused by meteorite impacts over millions of years. Past collisions involving large differentiated asteroids will have given rise to iron, stony, and stony-iron asteroids and meteorites.

Differentiation of their parent bodies is supported by the various kinds of meteorites found on Earth. For example an iron meteorite most likely originated from the core of a differentiated asteroid, and a stony-iron from the core-mantle boundary of such a body destroyed in a collision. Further support for differentiation and the early formation of differentiated asteroids comes from recent studies of a class of iron meteorites known as angrites. These indicate that their parent bodies had magnetic fields of the order of 20–40% of the present strength of that of Earth, for which you need an iron core, and that they formed in the first 3 million years of the Solar System's lifetime.

Binaries and Beyond

Many asteroids are binaries (double) or possess one or more small moons. Such binaries are formed when small, irregularly shaped 'rubble pile' asteroids have their rotation rates increased by the YORP (Yarkovsky–O'Keefe–Radzievskii–Paddack) effect – a result of the absorption and subsequent re-emission of solar radiation. (The YORP effect is a development of the Yarkovsky effect, which, as you will read in

Chap. 5, can cause changes in the orbit of an asteroid.) If the rotation period increases above 2–3 h material will be shed which may accumulate to form one or more small, nearby satellites. (243) Ida was the first asteroid to be found, in 1994 on images taken by the Galileo spacecraft, to have a natural satellite – Dactyl. In that same year the first binary system – two bodies of similar sizes orbiting one another – 1994 AW_1, was discovered. Using the Arecibo radio telescope in 2008, radar observations of NEO 2001 SN_{263} showed that this 2-km-diameter object had two moons, each approximately 600 m wide. Triple asteroids are not uncommon in the Main Belt, (45) Eugenia and (87) Sylvia being the first to be discovered.

In 2007 Asteroid (90) Antiope was found by the European Southern Observatory (ESO) to be not only a binary but a binary rubble pile (Fig. 4.4). How it got to be this way is unknown, but it might have been the result of a parent body spinning up, possibly by the YORP effect, as mentioned above, and breaking apart into two bodies, each between 80 and 90 m in diameter.

Although most EKBO binaries are closely coupled, a few are very widely separated – this separation varying from 4,000 to 40,000 km in the case of 1998 WW_{31}. How a pair of bodies could be so loosely coupled is somewhat of a mystery. One theory is that, when the EKB was still densely populated, two bodies could collide and coalesce in such a way as to lose most of their velocity. A third body could then capture this object into a highly eccentric wide orbit. Table 4.2 lists asteroids and dwarf planets with companions discovered up to January 2009.

Fig. 4.4. Asteroid (90) Antiope – binary rubble piles (artist's impression) (Credit: ESO).

Table 4.2. Asteroids and dwarf planets with companions

Group	Number with companions	Comments
NEOs	35	1 with 2 satellites
Mars Crossers	7	
Main Belt Asteroids	62	4 with 2 satellites
Jupiter Trojans	2	
TNOs, including Dwarf Planets	56	1 with 2 satellites and 1 with 3

Asteroid or Comet?

It is almost impossible to distinguish between an asteroid and a dormant or extinct comet on a CCD image. If the cometary give-away of a coma or tail(s) is missing there is nothing to tell them apart. Computing the orbit of the object may give clues to its make-up as asteroids tend to be in low inclination, low eccentricity prograde orbits, whereas comets are usually in high inclination, highly eccentric orbits. Their orbits may be prograde or retrograde, but the latter signifies that the object is definitely a comet. The Damoclids are an example of inactive comets. Astronomer David Jewitt says there is strong evidence (for example some have developed a coma) that these are inactive nuclei of the Halley family (20- to 100-year orbital periods) and long-period comets (orbital periods greater than 200 years).

Damoclids are not the only objects attempting to disguise themselves as asteroids. Asteroid 118401 has a comet-like dust tail and, together with comets P/2005 U1, 133P/Elst−Pizarro, P/2008 J2 (Belshore) and P/2008 R1 (Garradd), move in a typical Main Belt-like circular orbit. The tails of these three objects are most likely caused by subsurface water ice sublimating – turning directly from ice to gas. It could be that such bodies have brought large quantities of water to Earth when they were deflected out of their original Main Belt orbits, as described in Chap. 5.

Spectral Classification

Some objects, stars and hot gases, for example, produce their own light, and their spectra are referred to as emission spectra. On the other hand asteroids shine by reflected sunlight, and therefore their spectra are known as reflectance spectra. Their spectra are similar to the Sun's but with differences due to various minerals on the surface of the asteroid reflecting sunlight differently, and therefore the spectrum of a given asteroid will tell us something about the composition of its surface. Although composition is the major factor in determining the shape of spectra, other factors can influence this, namely, the reddening of spectra due to increased phase angle, darkening and reddening of the surface due to space weathering, size of particles making up the regolith, and surface temperature.

Figure 4.5 shows the spectra of a number of asteroids that are also listed under their spectral class (or type, as this is often referred to) in Table 4.3. The horizontal axis shows the wavelength of light in microns. Visible light, violet to red, falls in the range of 0.39–0.7 μm (390–700 nm), and 1.0 μm (1,000 nm) and beyond is in the near infrared part of the spectrum. The vertical axis indicates the amount of light reflected by the asteroid at specific wavelengths. This geometric albedo is the ratio between actual reflected light and what would be reflected by a perfectly white sphere, which reflects all incident light, e.g., its geometric albedo is 1.0. It is not uncommon on spectral diagrams of asteroids for the vertical axis to be 'adjusted' so that the albedo at a wavelength of 0.56 μm has the value of '1'.

Many asteroid spectra appear quite ordinary, but some do show strong features. For example the spectrum of (4) Vesta has a strong absorption feature centered on approximately 0.95 μm, indicating the presence of silicates. The slope of spectra at wavelengths greater than 0.55 μm indicates the presence or absence of materials such as iron, nickel, or organics that redden the surface of an asteroid. The redder the surface the greater is the upward slope from shorter to longer wavelengths.

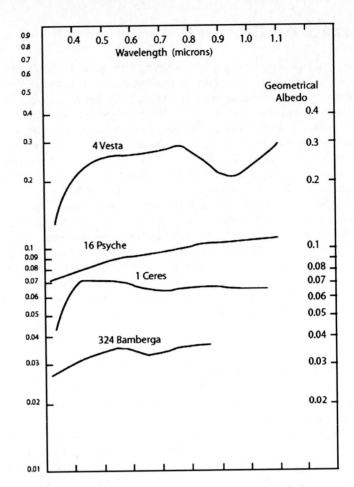

Fig. 4.5. Asteroid spectra (Credit: Steven K Croft from the original article by Chapman, Morrison, and Zellner).

(1) Ceres is an example of a relatively blue C-class asteroid, and it can be seen from its spectrum that it reflects more blue (shorter wavelength) light than red (longer wavelength).

An asteroid taxonomic (classification) system was first devised in 1975 by Chapman, Morrison, and Zellner. It included just three classes: C (dark carbonaceous objects which make up 75% of known asteroids); S (stony or silicaceous objects – 17% of known asteroids); and U for all others. The original system has been superseded by the (David J) Tholen and, more recently, the SMASS (Small Main-Belt Asteroid Spectroscopic Survey) classifications. The former included 14 types that the latter expanded to 22 as listed in Table 4.3.

The SMASS classification built on the Tholen classification but is based solely on the presence or absence of absorption features in the visible part of the spectrum. In many cases the two classifications are the same, but the Tholen C and S classes were subdivided in the SMASS taxonomy. This table gives only the briefest summary of what is quite a complex subject. Readers should note that not all asteroids are yet classified, and some do change type as more observations are made. As noted in the table the proportion of different classes of asteroids varies across the width of the Main Belt.

Table 4.3. SMASS (Small Main-Belt Asteroid Spectroscopic Survey) classifications

SMASS class	Tholen class	Description	Examples (by SMASS class)
B	B, C, F, G	Uncommon class of carbonaceous asteroid found mainly in the outer Main Belt, peaking in abundance at 3 AU from the Sun. Spectra show the presence of clays, carbon, and organics	(2) Pallas, (431) Nephele (704) Interamnia (25143) Itokawa
C, Cb, Cg, Cgh, Ch		Carbonaceous (chemical composition is similar to the Sun but not including volatiles such as hydrogen and helium). Relatively blue in color with spectra similar to carbonaceous chondrite meteorites. Most common class of asteroid forming a large proportion of the outer Main Belt population	(1) Ceres (10) Hygeia (19) Fortuna (45) Eugenia (90) Antiope (253) Mathilde (324) Bamberga (379) Huenna (2060) Chiron
A	A	Uncommon inner Main Belt asteroids, most likely to be from the mantle of a differentiated parent body. Spectra indicate presence of olivine	(246) Asporina (289) Nenetta
Q	Q	Relatively uncommon inner Main Belt asteroids. Spectra indicate presence of olivine and pyroxine and are similar to ordinary chondrite meteorites	(1862) Apollo
R	R	Moderately bright, relatively uncommon inner Main Belt asteroids. Spectra indicate olivine, pyroxene, and plagioclase present	(349) Dembowska
K	S	Silicaceous or stony and reddish in color. Relatively uncommon Main Belt asteroids with a featureless spectrum found during studies of the Eos family. Spectra similar to stony-iron meteorites indicate presence of olivine and orthopyroxene	(221) Eoa (233) Asterope
L, Ld		Relatively uncommon asteroids with featureless spectra. Ld-class asteroids are similar to the L-class but redder	(387) Aquitania (728) Leonisis (?)
S, Sa, Si, Sk, Sr, Sq		Moderately bright, stony, chondritic asteroid class dominating the inner Main Belt, peaking at 2.3 AU but decreasing in proportion towards the outer Main Belt	(6) Hebe, (15) Eunomia, (433) Eros
X, Xe, Xc, Xk	E, M, P	Metallic asteroids mostly located at 2 AU (Tholen E-class) and 4 AU (Tholen P-class). Spectra indicate presence of troilite (iron sulphide), enstatite, and hypersthene — the latter two contain magnesium and hypersthene, also iron. Anhydrous silicates and organics may also be present on the surface of P-class. M-class asteroids have spectra similar to almost pure iron–nickel	(16) Psyche, (44) Nysa (64) Angelina (87) Sylvia (216) Kleopatra (2867) Steins
T	T	Rare, low albedo (dark) inner Main Belt asteroids of unknown composition with featureless, moderately red spectra	(114) Kassandra (?)
D	D	Reddish spectrum of these outer Main Belt objects (proportion greatest at 5.2 AU) may indicate presence of organic rich carbon and anhydrous silicates and possibly ice. Spectra similar to the Tagish Lake meteorite — a carbonaceous chondrite	(624) Hektor (944) Hidalgo Martian moons have similar spectra and may be captured D-class asteroids
O	–	Spectra have strong absorption feature at wavelengths longer than 0.75 μm	(3628) Boznemcova
V	V	Similar to S-class but containing a form of pyroxene known as augite. Spectra a close match to eucrite meteorites	(4) Vesta

An example of an S-class asteroid is (433) Eros. The image in Fig. 4.6 was obtained by the NEAR spacecraft, which visited this asteroid and actually landed on it in February 2001.

Meteorites have been linked to various groups and types of asteroids. In 1964 the orbits of several ordinary chondritic meteorites were determined by simultaneously photographing their trajectories. It was shown that the aphelion of their

Fig. 4.6. S-class asteroid (433) Eros (Credit: NASA/Johns Hopkins University Applied Physics Laboratory).

orbits fell within the asteroid belt and that therefore they might be pieces of asteroids. Spectroscopic analysis indicates that such meteorites may have originated from S-class asteroids. Other meteorite-asteroid links are Eucrites (a subset of achondrites) and (4) Vesta, Aubrites (also a subset of achondrites) and (44) Nysa, Irons and (16) Psyche, and Stony-irons and (246) Asporina.

Asteroids that may once have been part of a large, differentiated object are V-class (formed from lava that once flowed on or near the surface), A and R-classes (previously part of the mantle), S-class (core-mantle boundary, mantle, lower crust), and X-class (core).

Identifying the parent body of a subset of chondritic meteorites known as ordinary chondrites proved problematic. These were thought to originate from S-class asteroids, for example (6) Hebe, which are found in large numbers in the inner Main Belt. Spectral analysis of such asteroids indicated a lack of silicates and a surplus of iron compared with the associated meteorites. The difference is thought to be due to space weathering, whereby bombardment by the solar wind causes minerals in the surface layer of the asteroid to evaporate and re-condense as a film of iron particles.

Opposition Effect

As an asteroid approaches opposition it will usually brighten by more than might be expected – not unusually by 0.3–0.5 magnitudes. This opposition effect and how to measure it will be explained in more detail in Chap. 14. Suffice it to say that it varies with asteroid type and albedo. Low albedo equals little opposition effect and can tell us something about the nature of the surface of the asteroid and the way it scatters incident light.

What of EKBOs?

The faintness of objects in the Edgeworth–Kuiper Belt makes it extremely difficult even with the largest telescopes to obtain spectra of these objects. Dave Jewitt obtained a spectrum, Fig. 4.7, of (50000) Quaoar using the 8-m Subaru telescope on Mauna Kea.

The black line shows Subaru data and the red line is a spectrum of water ice plotted for comparison. Broad minima at 1.5 and 2.0 μm indicate water ice on the surface. A sharper minimum at 1.65 μm shows that the ice is crystalline rather than amorphous. The crystalline nature of the ice indicates that the temperature on Quaoar must have been above −160°C at the time of its formation. One, rather speculative, scenario for the formation and continued existence of ice on the object's surface goes like this:

1. Ice in the interior of Quaoar is melted by radioactive decay.
2. The presence of ammonia lowers the melting point of the ice and the resulting water percolates onto the surface.
3. The solar wind and cosmic rays are prevented from turning the crystalline ice into amorphous ice by micrometeorite impact or the addition of fresh crystalline ice from the interior.

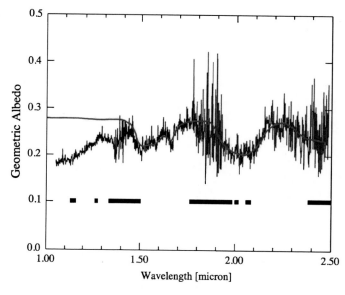

Fig. 4.7. Reflection spectrum of EKBO (50000) Quaoar (Credit: Dave Jewitt).

Unusual or anomalous cosmic rays reaching Earth can tell us something about the Edgeworth–Kuiper Belt. Researchers believe that the solar wind dislodges particles and atoms from EKBOs, which eventually find their way back to us after a journey to the outer reaches of the Solar System.

Dwarf Planets and Their Moons

As mentioned in Chap. 2 dwarf planets are bodies that are large enough for their own gravity to pull them into an approximately spherical shape. That is, they are larger and more rounded than the average asteroid but not so large as a fully fledged planet. They do span the whole age of discovery of the smaller bodies in the Solar System with (1) Ceres being the first asteroid to be discovered and what we now call dwarf planets being found in more recent times in the far regions of the Solar System.

There are currently five dwarf planets. These, and their moons, are described in more detail below and in Tables 4.4 and 4.5. All such objects have a diameter of close to or exceeding 1,000 km, but no minimum size for an asteroid to be 'upgraded' to dwarf planet status has yet been given by the International Astronomical Union. There may be as many as 40 or 50 known asteroids that could

Table 4.4. Physical properties of the known dwarf planets

	(1) Ceres	(134340) Pluto	(136108) Haumea	(136199) Eris	(136472) Makemake
Provisional designation	–	–	2003 EL$_{61}$	2003 UB$_{313}$	2005 FY$_9$
Diameter (km)	909.4 (polar) × 974.6 (equatorial)	2,390	996 (polar) × 1,960 (equatorial)	2,600 ± 100	1,500 + 400/−200
Mass (×10^{21} kg)	0.94	12.5	4.2 ± 0.1	1.67 ± 0.02	~4
Density (g/cc)	2.08	1.75	2.6–3.3	~1.8	~2
Albedo	0.09	0.5–0.7	0.7 ± 0.1	0.86 ± 0.07	0.78
Atmosphere	None	Transient	Unknown	Transient (?)	Transient (?)
Mean surface temp (K)	~167	~43	<50	42.5	?
Rotational period (h)	9.07	153.3	3.9	>8 (?)	?
Axial Tilt (°)	~3	122.53	?	?	?

Table 4.5. Satellites of the dwarf planets

	Distance from planet (km)	Orbital period (days)	Diameter (km)	Mass (kg)
(134340) Pluto				
Charon	19,571 ± 4	6.4	1,207 ± 1.5	1.52 ± 0.06 × 10^{21}
Hydra	64,749	38.2	60–168	5 × 10^{16} − 2 × 10^{18}
Nix	48,708	24.9	46–136	5 × 10^{16} − 2 × 10^{18}
(136108) Haumea				
Namaka	39,300^a	18	~170	~2 × 10^{18}
Hi'iaka	49,500 ± 400	49.1	~310	~2 × 10^{19}
(136199) Eris				
Dysnomia	37,370 ± 150	15.8	100–300	?

^{a} Distance from planet when observed

achieve dwarf planet status, and there may be as many as 2,000 such objects in the Edgeworth–Kuiper Belt – 2007 OR$_{10}$ being the latest and seventh largest asteroid to be discovered.

(1) Ceres

Ceres, shown in Fig. 4.8, is the largest Main Belt asteroid and is slightly oval in shape, measuring 975×909 km.

In classification terms it was originally designated a planet, then demoted to an asteroid but recently partially restored to its former glory by being included in the dwarf planet category. Its size and mass have allowed it to achieve this nearly spherical shape; in technical terms it is a gravitationally relaxed spheroid. Those with excellent eyesight and access to a dark site may be able to observe Ceres with the unaided eye. At its brightest it can reach magnitude 7 (the only other asteroid likely to be seen in this way is (4) Vesta). It is believed that Ceres is a differentiated body with a rocky core and icy mantle between 60 and 120 km thick. This mantle, made up of water ice and hydrated minerals such as carbonates and clays, may contain more water than is found on Earth. There are some indications that Ceres may have a tenuous atmosphere and frost on its surface. We should learn much more about this dwarf planet when the *Dawn* spacecraft reaches it in 2015.

Ceres is unusual in that it is the only dwarf planet in the Main Belt. However the Nice model, described in Chap. 5, proposes that all carbonaceous, or C-type, asteroids in the Main Belt and Jupiter Trojan population were originally formed between 5 AU and the outer edge of the Edgeworth–Kuiper Belt (possibly 50–100 AU). The dwarf planet (1) Ceres may therefore be related to the other dwarf planets, which are found in the outer reaches of the Solar System.

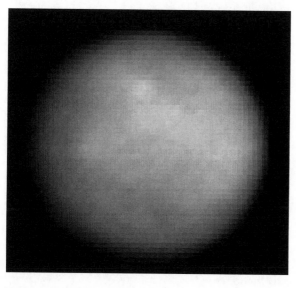

Fig. 4.8. The dwarf planet (1) Ceres (Credit: Space Telescope Science Institute).

(134340) Pluto

Entire books – see the Resources section at the back of this book – have been written on the subject of Pluto and its discovery by Clyde Tombaugh on February 18, 1930, so only the briefest description is included here.

Pluto was always an oddball planet with its highly eccentric and highly inclined orbit. Discovery of more large objects in the outer regions of the Solar System finally gave it a home to go to. In recent times there has been much, perhaps too much, discussion on its reclassification as a dwarf planet. Pluto is a differentiated body thought to have a large rocky core covered with a layer of frozen water and other icy materials such as nitrogen, plus small amounts of methane and carbon dioxide. We should learn much more about Pluto and its three known satellites when the *New Horizons* spacecraft arrives there also in 2015. An image of Pluto obtained by this spacecraft on October 6, 2007, is shown in Fig. 4.9.

Pluto's distance from the Sun varies considerably. At its closest, perihelion, it is 39.4 AU distant, closer than Neptune, whereas at its furthest, aphelion, it is 49.3 AU away. However, being locked in a 3:2 resonance with its neighbor means a collision between the two is not possible. As a result of this varying distance from the Sun the frozen nitrogen plus small amounts of methane and carbon dioxide on its surface vaporize at and near perihelion, forming a thin atmosphere. In early 2009 observations using the European Southern Observatory's Very Large Telescope showed that the atmosphere, at 180°C, is 40°C hotter than the surface. It is believed that this may be due to the presence of pure methane patches or a methane-rich layer covering Pluto's surface. As the dwarf planet moves towards aphelion its atmosphere gradually shrinks as the gases freeze onto its surface. Because the incident sunlight is used to defrost the nitrogen the surface of Pluto is actually colder than that of its moon

Fig. 4.9. First detection of Pluto by the *New Horizons* spacecraft (Credit: NASA/Johns Hopkins University Applied Physics Laboratory/ Southwest Research Institute).

Charon by about 10°C. The reddish color of Pluto is believed to be due to deposits of tholins on its surface. Tholins are complex organic compounds produced by solar ultraviolet irradiation of, for example, methane or ethane.

Stellar occultations (explained in Chap. 15) have given us more information on Pluto's atmosphere. They indicate a thickening of the atmosphere and a slight warming of the surface up to at least 10 years after perihelion.

Pluto's three moons are locked in a 12:3:2 resonance of orbital periods. This suggests that they were formed at the same time, possibly by the breakup of a larger body due to a collision. Charon is large compared to its parent planet – so large in fact that the Pluto-Charon system is often referred to as a binary planet.

(136108) Haumea

This dwarf planet was discovered in 2004 by a team of astronomers led by Mike Brown at Caltech. It has a rather unusual ellipsoid shape (see Fig. 4.10), unique among the known EKBOs in that it is some way from 'nearly round,' as required by the IAU definition for dwarf planets – more the shape of a rugby ball or an American football.

It is believed that Haumea was once part of a much larger body, 50% rock and 50% ice, about the same size as Pluto, broken up by a collision with another EKBO. This collision led to the high rotation rate of the parent body and knocked off most of an outer icy layer that then formed the two moons, Namaka and Hi'iaka, and a number of other nearby EKBOs. This theory is supported by measurements of the surface composition of Hi'iaka, which shows it to be possibly pure, water ice. Haumea itself is thus almost 100% rock, with just a very thin layer of ice on the surface. This ice is crystalline, indicating that it froze slowly. In this region of the Solar System ice should have formed quickly and be of the amorphous variety. Why the ice on the surface of Haumea is crystalline is still a mystery (but see the section 'What of EKBOs?' above for a possible explanation).

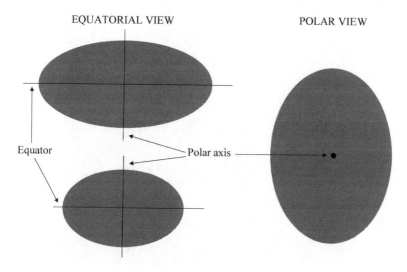

Fig. 4.10. Unusual shape of dwarf planet (136108) Haumea (Diagram by the author).

(136199) Eris

This dwarf planet was discovered on October 21, 2003, by the same team that found Haumea: Mike Brown, Chad Trujillo, and David Rabinowitz. The team used the 60-year-old, 48-in. (1.2-m) Samuel Oschin Telescope on Mt. Palomar (see Fig. 4.11). Readers can compare this instrument with those used by amateurs described in Chap. 7. It just goes to show that old equipment (and old astronomers!) can still get good results.

Eris, currently the largest known dwarf planet, is another with a highly eccentric orbit. At aphelion it is 97.5 AU from the Sun, but at perihelion only 38.3 AU. Objects in such orbits are classified as scattered disk objects. It is also a plutoid, as are Makemake and Pluto itself – see Chaps. 2 and 3 for more on plutoids and plutinos. Its spectrum is very similar to that of Pluto, indicating that the surface is covered with frozen methane, plus a small amount of nitrogen ice that makes the object grayish in appearance. The interior of Eris is also likely to be similar to Pluto – a mixture of rock and ice. At and close to perihelion, the sunlit surface may be warm enough for the frozen gases to vaporize and thus, for this part of its orbit, Eris may possess a very thin atmosphere. Near aphelion, where Eris is now, it displays a uniform, almost white, surface reflecting approximately 90% of the Sun's light. It is believed this is due to most of its atmosphere freezing onto the surface at this distance from the Sun. The surface temperature on Eris is approximately −240°C at aphelion and −170°C at perihelion.

Fig. 4.11. Samuel Oschin Telescope used to discover (136199) Eris (Credit: Caltech/Palomar Observatory).

(136472) Makemake

Another dwarf planet discovered by Mike Brown and colleagues in March 2005. Not so many years ago it was thought that imaging TNOs was far beyond the capabilities of even the experienced amateur. Today, with the right equipment, it is quite an easy task, as will be demonstrated in Chaps. 10 and 11. Figure 4.12 shows the movement of Makemake over a 5-day period in a series of images obtained by the author using the Skylive Robotic Telescope at Grove Creek Observatory, New South Wales, Australia.

As is quite common for these recently discovered EKBOs, now classified as dwarf planets, Makemake is extremely bright. As in the case of Eris this may be due to the atmosphere freezing onto the surface. Like many objects in the far reaches of the Solar System Makemake is red in color, its spectrum indicating the presence of methane and ethane but no nitrogen or carbon monoxide as are found on Pluto. The methane absorption lines in the spectrum are broad compared to other Solar System objects, indicating that the solid methane is in the form of 1-cm size 'hailstones.'

The next chapter portrays how and where these small inhabitants of the Solar System were formed and what caused them to arrive at their present locations.

Fig. 4.12. Dwarf planet (136472) Makemake imaged on Jan 31, Feb 3, and Feb 4, 2009 (Images by the author).

Chapter 5

Origins and Evolution

Having described in the previous chapters the locations and structure of asteroids and dwarf planets, this part of the book shows how they came to be where we find them today and includes:

- The birth of the Solar System
- How asteroids formed
- Where near-Earth asteroids come from
- Why are there asteroids but no planets between Mars and Jupiter?
- Planetary satellites
- The origin of Trojan asteroids
- Unpredictable Centaurs
- The Edgeworth–Kuiper Belt (aka trans-Neptunian objects, or TNOs) – knowns and unknowns

The Birth of the Solar System

About 15 billion years ago there was a 'Big Bang.' As far as we asteroid observers are concerned, nothing much of interest occurred for the next 10 billion years or so. Then, about 4½ billion years ago, as one theory has it, the shockwave from another 'Not-Quite-So-Big Bang,' a nearby supernova, caused a cloud of gas and dust to collapse in on itself. Such a nearby supernova would also have created and added heavy elements to the collapsing nebula. Rising density and temperature of the central region eventually reached the point where nuclear reactions, the conversion of hydrogen to helium, began, and our Sun was born. The rest of the material settled into a disk circling the new-born star.

The planets and asteroids were formed by the gradual agglomeration, due to gentle collisions, of particles into centimeter- and then meter-sized bodies. As these small bodies increased in size they were able to gravitationally attract other material and continue to grow. Close to the Sun, where temperatures were high, most of the icy material was evaporated and blown outwards. The dust grains left behind eventually came together to form the four inner rocky planets: Mercury, Venus, Earth, and Mars. Further out, beyond the 'snow line,' at around 3 AU from the Sun (where temperatures were much lower), dust, ice, and gas coalesced to form the gas giants: Jupiter, Saturn, Uranus, and Neptune.

R. Dymock, *Asteroids and Dwarf Planets and How to Observe Them*,
Astronomers' Observing Guides, DOI 10.1007/978-1-4419-6439-7_5,
© Springer Science+Business Media, LLC 2010

How the Asteroids Formed

Not all the original material formed itself into large planets. Some remained as small rocky bodies (asteroids) and 'dirty snowballs' (comets). Asteroids formed extremely quickly, as recent observations by a team of scientists led by the University of Maryland show. Their research has identified three asteroids with an age of 4.55 billion years.

Computer simulations by Stuart J. Weidenschilling and the Cosmic Dust Agglomeration Experiment on the Space Shuttle *Discovery* run by Jurgen Blum and colleagues support the following process. First, gentle collisions allowed small, micron-sized or less, dust particles in the disk to stick together to form fluffy aggregates (see Fig. 5.1). In turn these combined into centimeter-sized, low density lumps and remained 'glued' together. These small objects grew both by attracting more dust particles and combining with one another. It wasn't all one-way traffic, though, as some collisions were violent enough to shatter the smaller agglomerations back into dust.

As we know from observations, asteroids vary in size from a meter or two to several hundred if not thousands of kilometers in diameter. The larger bodies, a few hundred kilometers or larger, were able to differentiate to form an iron core with a surrounding rocky mantle and crust.

Where Near-Earth Asteroids Come from

Many of those meter- to kilometer-sized bodies located in the inner Solar System at the time of its formation were removed by collisions with the newly forming planets or by being ejected from the Solar System by gravitational interaction with the larger bodies.

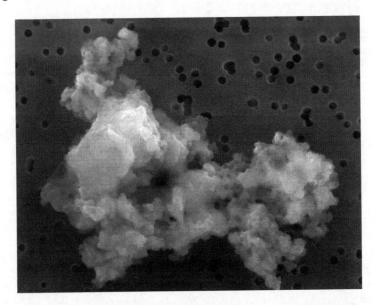

Fig. 5.1. Interplanetary dust retrieved by a U2-type aircraft. Image size is approximately 10 × 8 μm. (Credit: NASA) (*For aviation buffs, the ER-2 is NASA's version of the U2-C model. NASA has since acquired and used the U2-R or TR-1 model, but has retained the ER-2 descriptor. The newest ER2 (U2-R) was built and delivered in 1989 and represents one of NASA's youngest aircraft.).

The life of an asteroid as an NEO is relatively short – 2 to 6 million years according to William F. Bottke Jr. – before it is lost due to an impact with Earth, the Moon, or other inner planet or by being ejected from the inner Solar System. Cratering studies show that the impact rate has remained unchanged for the last 3 billion years and therefore the number of NEOs must have been reasonably constant over the same period. That being the case, new NEOs need to be created at a rate of approximately one every 70,000 years.

Most NEOs have their origins in the Main Belt. The collision of two such asteroids could result in one or both, or fragments of either, entering one of the unstable regions known as Kirkwood Gaps (see below).

Another mechanism by which asteroids with diameters of 20 km or less have their orbits changed gradually and thus moved into the unstable regions of the Main Belt is known as the Yarkovsky effect. It is believed that this mechanism is the dominant of the two.

As asteroids rotate they absorb and then reradiate the Sun's heat. This re-emission acts to change the orbit of the asteroid, as shown in Fig. 5.2. In the case of asteroid Y, which is spinning in the same direction in which it orbits (prograde), the force acts to speed up the asteroid, thus moving it outwards from the Sun. The force on asteroid X, which is spinning in the opposite direction to which it orbits (retrograde), acts in the opposite direction, slowing it down and causing it to spiral inwards towards the Sun. The forces are very small; radar observations of asteroid (6489) Golevka showed that this 0.5-km-diameter asteroid had shifted its orbit by only 15 km in the 12 years from 1991 to 2003.

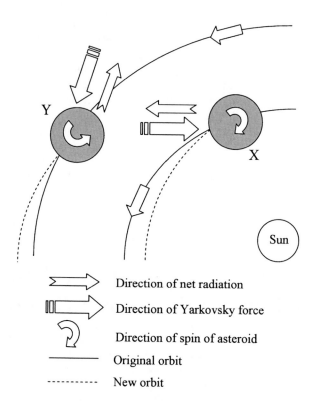

Fig. 5.2. How the Yarkovsky effect changes the orbit of an asteroid (Diagram by the author).

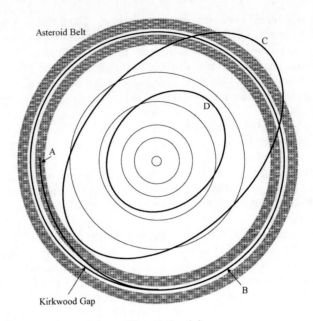

Fig. 5.3. The path from Main Belt to near-Earth asteroid (Diagram by the author).

Having made its way into, for example, the 3:1 Kirkwood Gap (A to B) in Fig. 5.3, the orbit of an asteroid becomes strongly influenced by Jupiter's gravity. The greatest effect of such a perturbation is to increase the eccentricity of its orbit so that it eventually became a Mars crosser (C). Further perturbed by Mars, its new orbit would then bring it in to the inner Solar System as an Apollo or Aten asteroid (D).

A typical example of such a life story relates to the death of the dinosaurs 65 million years ago. Research by a team of U. S. and Czech astronomers suggests that the asteroid that struck the Yucatan peninsula was most likely part of the Baptistina family. The comparatively young crater Tycho, on the Moon, was also quite likely to have been formed by an asteroid from this same group. The parent body was broken up, into about 140,000 bodies larger than 1 km in diameter, by a collision 160 ± 20 million years ago. Fragments propelled by the Yarkovsky effect, drifted into one of the Kirkwood Gaps and were then propelled into Earth-crossing orbits. Research has shown that, due to bombardment by members of the Baptistina family, impacts on Earth and the Moon have more than doubled over the past 100–150 million years.

Why Are There Asteroids and No Planets Between Mars and Jupiter?

The vast majority of known asteroids lie within this belt between approximately 2.0 and 3.3 AU from the Sun. The large numbers of small bodies with low relative velocities allowed smaller asteroids to grow into larger ones by collision.

Some mechanism caused these circumstances to change. Asteroids became fewer in number and orbital eccentricities and inclinations increased, leading to an increase in their relative velocities. The reason for the change could have been perturbations by Jupiter as it moved closer to the Sun, as explained in 'The Edgeworth–Kuiper Belt: Knowns and Unknowns' below, or the passage of one or several massive objects through the asteroid belt. Collisions at higher velocities would cause asteroids to fragment rather than coalesce.

The Main Belt, and the Jupiter Trojan groups, may have been partially populated by asteroids, particularly the carbonaceous or C-type, originating between 5 AU and the outer edge of the Edgeworth–Kuiper Belt and arriving in the outer Main Belt during the time of the Late Heavy Bombardment (LHB) 3.8 million years ago.

Why Are There Gaps in the Main Belt?

The distribution of asteroids varies considerably across the width of the belt. Where the orbital period would be a simple fraction of that of Jupiter, for example 1/2 or 1/3, we find 'asteroid free zones,' known as Kirkwood Gaps (Fig. 5.4). These gaps are an example of resonances, usually written as 2:1 or 3:1 – a situation in which orbiting bodies, asteroids in this case, are subjected to regular gravitational disturbances by another, in this case Jupiter. Under the influence of regular tugs by Jupiter most asteroids have been cleared out of these gaps. Beyond 3.5 AU, for reasons not fully understood, resonances act to keep asteroids in their present orbits rather than perturb them into different paths.

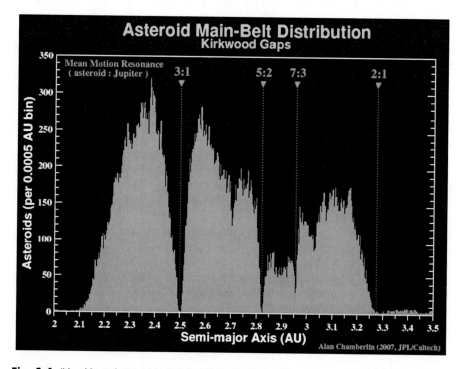

Fig. 5.4. Kirkwood Gaps in the Main Belt (Credit: NASA/JPL/Caltech/Alan Chamberlain).

The Edges of the Main Belt

The position and shape of the inner edge of the Main Belt is the result of a secular resonance. In this case the precession of the nodes (see Chap. 2 for more on nodes) of an asteroid's orbit is in resonance with the precession of the nodes of Saturn's orbit. This resonance is quite powerful and will act to increase the eccentricity of the orbit of an asteroid in this region and eventually cause it to be removed from the Main Belt, as described in the section above, 'Where Near-Earth Asteroids Come from.'

The possible inward migration of Jupiter in the formative years of the Solar System (see 'The Edgeworth–Kuiper Belt: Knowns and Unknowns' below) cleared small bodies from beyond the present outer edge of the Main Belt.

How Asteroid Families Came to Be

In 1918 statistical studies by K. Hirayama showed that almost half the Main Belt asteroids could be grouped by combinations of certain of their orbital elements, e.g., semi-major axis, inclination, and eccentricity. These Hirayama families are the result of break-ups, by collisions perhaps occurring only once in 10 to a 100 million years, of larger, parent asteroids. It should be noted that Hirayama wasn't the first to use the term 'family' in this respect, as William Monck, a founding member (no. 12) of the British Astronomical Association, referred to the existence of small groups of asteroids by that name back in 1888.

One such collision happened recently (in astronomical terms), just under 6 million years ago. David Nesvorny and his colleagues identified a group of asteroids, the Karin cluster, with similar orbital elements and traced their orbits backwards in time to the point where they merged. The largest of the 39 fragments of the original body, approximately 19 km in diameter, is (832) Karin. It is believed that this cluster is part of the larger Koronis family.

Another method of identifying families, particularly where it is difficult to match orbital characteristics, is by color. Members of a family of asteroids should have the same color as their parent body. These colors are very subtle and need something like the Sloan Digital Sky Survey's 2.5-m telescope and its range of filters to differentiate them. Zeljko Ivezic and his colleagues used Sloan data to identify families by color. They concluded that at least 90% of Main Belt asteroids could be grouped into families – considerably more than Hirayama suggested, but then he didn't have the use of the Sloan telescope!

There are exceptions to the above 'rule' in that some families are composed of different types of asteroids, indicating that their parent bodies had differentiated, e.g., were composed of a metallic core surrounded by layers of rocky material. The Vesta family is an example of such a group, although in this case the parent body was not totally destroyed in the collision that created this family.

Asteroid families are not confined to the Main Belt. Mike Brown and colleagues have discovered a number of objects near EKBO 2003 EL_{61} with similar surface properties and orbital elements. They believe that these objects are fragments of 2003 EL_{61} which was involved in a collision at the same time as Earth was forming 4.5 billion years ago.

Planetary Satellites

An asteroid approaching a planet cannot be captured into orbit by that planet's gravity alone. (A spacecraft will use retro-rockets, but not too many asteroids are fitted with these!) One possibility is that, when the planets were forming, they were surrounded by gas disks or much larger atmospheres than exist today. A gas disk or extended atmosphere could have slowed down an approaching asteroid and captured it into orbit. Many more potential moons would have been lost than captured; too steep an angle of approach and the asteroid would burn up, too shallow and it would skip off the atmosphere back into space. Another possibility is that one body of an approaching binary was captured and the other escaped.

Yet another scenario has been put forward for the existence of Mars's satellite Phobos. Preliminary density calculations based on data from the ESA *Mars Express* mission suggest that this moon is a highly fractured body – a rubble pile. It is possible that this material was blasted off the surface of Mars by a large meteorite impact and subsequently coalesced into the moon. This is not beyond the realms of possibility, as analysis of data obtained by the NASA *Mars Reconnaissance Orbiter* has identified the largest impact crater found anywhere in the Solar System. Our own Moon was probably formed in a similar way, but, in that case, the much larger 'rubble pile' coalesced to form a solid body.

The Origin of Trojan Asteroids

As mentioned later in this chapter, one of the major influences on asteroid evolution was the time when Jupiter and Saturn were in a 2:1 resonance. Computer simulations by Morbidelli and his colleagues showed that Jupiter could capture large numbers of asteroids into Trojan orbits. These objects had previously resided in a belt just beyond the orbit of Neptune (see Fig. 5.5), and their orbits were significantly altered by the changing orbits of the gas giants. The computer model was supported by observations in 2004/2005 using the Keck II telescope. These showed that Jupiter Trojan asteroid (617) Patroclus had the same composition (ice with a coating of dirt) as comets and, more importantly, small EKBOs.

Unpredictable Centaurs

It is likely that most Centaurs, (2060) Chiron and (5145) Pholus, for example, were originally Edgeworth–Kuiper Belt objects. Chiron could have had its origin in the Oort Cloud, as it may actually be a comet that is outgassing and may have arrived in its present position under the gravitational influence of Neptune. Under the continuing influence of Neptune and the other giant planets, Uranus, Saturn, and Jupiter, its orbit is unstable. It, and other Centaurs, could evolve inwards possibly leading to a collision with one of the planets, or outwards even, being ejected from the Solar System.

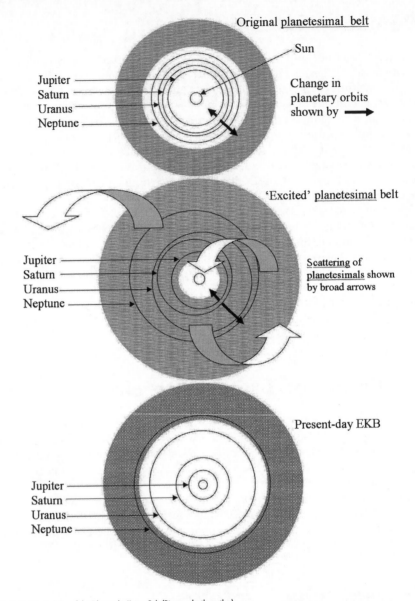

Fig. 5.5. Evolution of the Edgeworth–Kuiper Belt (Diagram by the author).

The Edgeworth–Kuiper Belt: Knowns and Unknowns

According to the Nice model, developed by Rodney Gomes, Hal Levison, Alessandro Morbidelli, and Kleomenis Tsiganis in 2004, the giant planets formed much closer to the Sun. They were originally between 5 and 15 AU from the Sun compared with the 5–30 AU where they reside today. Beyond the giant planets lay a wide belt of smaller bodies that, although large in number, did not encounter one another frequently enough to allow planets to be built. During the period of planet formation these bodies grew steadily, by accretion of smaller planetesimals, to sizes of, typically, hundreds of kilometers in diameter.

Those of a nervous disposition should avert their eyes from the text below and Fig. 5.5, as what happened next was violent in the extreme. The gravitational forces exerted by the giant planets on the belt of smaller bodies caused their nearly circular orbits to become more eccentric. Collisions between these objects, which had previously been gentle enough to enable the objects to coalesce, now caused them to fragment. At the same time some were flung outwards and even completely out of the new-born Solar System. Others came inwards and bombarded both the terrestrial planets and the gas giants. Earth was nearly destroyed and the Moon formed by such an encounter. In a way this was fortunate for us, as the Moon stabilizes the tilt of Earth's axis at 23½°, leading to a more benign climate.

Not only did the gas giants affect the orbits of the myriad of planetesimals, but the reverse was also true. Initially the giant planets resided in the region between approximately 5.5 and 14 AU from the Sun. While Saturn, Uranus, and Neptune moved slowly outwards, Jupiter moved slightly closer to the Sun until the orbital period of Saturn was exactly twice that of Jupiter. This 2:1 resonance, two orbits of Jupiter for each one of Saturn, caused the orbit of Saturn to change from being roughly circular to quite eccentric, causing it to make close passes of both Uranus and Neptune. These two planets were thrust into the zone of planetesimals, grossly altering the orbits of all concerned (Uranus and Neptune may even have switched positions).

At around 3.8 billion years ago, the inner planets were subjected to a rain of large objects, the Late Heavy Bombardment (LHB), the results which are still visible on the Moon. Research in 2009 on maria by a team from the Niels Bohr Institute led by Uffe Grae Jørgensen showed that the impactors were comets rather than asteroids (supporting a distant origin for these objects beyond what is now the Main Belt and adding further credence to the Nice model). Saturn, Uranus, and Neptune moved outwards to their present positions at 9.5, 19.2, and 30.1 AU, respectively, while Jupiter moved marginally inwards to its current location at 5.2 AU from the Sun.

In all probability 99% of the objects originally in the planetesimal belt beyond Neptune have been lost, and the remaining ones cajoled into what we now know as the Edgeworth–Kuiper Belt and even as far out as the Oort Cloud. The changing orbits of the gas giants flung the objects in this region every which way. Not only were objects dispatched to other parts of the Solar System, but many were destroyed by collisions in the process. The resulting dust particles would easily have been blown into interstellar space by the pressure of the Sun's radiation. The dusty disks observed around other stars may be the result of similar activity.

Neptune exerts considerable influence over the ongoing evolution of the present-day EKB. Computer modeling suggests that the plutinos and classical EKB asteroids are in stable, resonant orbits with Neptune and are the survivors of that original, much larger population. The dwarf planet Pluto and the plutinos have orbital periods that are 1.5 times that of Neptune; they complete two orbits for every three made by that planet. Such relationships lead to stable orbits, and these objects are described as being in a 2:3 mean motion resonance with Neptune. Similarly various classical EKBOs are in 5:3, 7:4, or 2:1 mean motion resonances with Neptune.

Those asteroids not in resonant orbits came close to Neptune and were ejected from that region to form the population of scattered disk objects. It may be that their orbits were further influenced by one or more large bodies, of planet or dwarf planet size, which existed in or passed through the region at that time. Such a body may still exist if Patryk Lykawa and his colleagues are correct. (Remember Planet X mentioned in Chap. 3?) At perihelion, as close as 35 AU from the Sun in some cases, they come near enough to Neptune to be influenced by that planet's gravity, which gradually causes their orbits to become both more eccentric and more inclined to the ecliptic. Eccentricity, inclination, and the longitudes of perihelion and the rising node can all oscillate slowly if the object in question is in a resonance with another body. These variations were first described by Yoshide Kozai, who gave his name to them.

Close encounters between SDOs also tend to increase the eccentricity of their orbits. At these large distances from the Sun their motion might also be influenced by a passing star or the passage of the Solar System through the central plane of the galaxy or a giant molecular cloud. If the Solar System had formed in a stellar cluster there may have been more encounters in the past than are likely at present or in the future. These would tend to increase both the inclination and eccentricity of the objects' orbits. Mike Brown and colleagues suggest that the highly eccentric orbit of inner Oort Cloud object 2003 VB_{12} is the result of one such encounter. SDOs will eventually become Centaurs, where their orbits are easily influenced by the gravitational tugs of the gas giants, as described earlier. There is also an outside chance that our Solar System may have exchanged a large number of asteroids and/or planetesimals with a passing planetary system.

Is there an outer edge to the Edgeworth–Kuiper Belt? A difficult question to answer, detection being difficult even with the most modern automated surveys. Objects at distances of 50–100 AU from the Sun are very faint and move very slowly against the background stars. There seems to be no good reason why more objects shouldn't exist among and beyond those already discovered, so we will just have to wait for the detection abilities of hardware and software to improve, as they surely will. The remoteness and faintness of these objects does not deter amateur astronomers from initiating projects to search for them. One such endeavor is Eamonn Ansbro's 'A survey telescope for detecting Edgeworth–Kuiper Belt objects' – read about it in Chap. 11.

As mentioned earlier in this chapter NEOs are those bodies that find their way, mostly from the Main Belt, into Earth-crossing orbits. The next chapter describes the possible effects of an asteroid impact, what is being done to discover and track potential impactors, scenarios for impact prevention, and the significant role played by amateur astronomers in this work.

Chapter 6

Impact?

This chapter is not so much about the actual impacts but concentrates on how potential impactors are so defined and what might be done to deflect or destroy them. Since amateur astronomers have an important role to play here, it seems necessary to describe the state of play and make the point that the rules of this particular game are far from complete.

Near-Earth Objects (NEOs)

As mentioned in Chap. 3, near-Earth objects come in several guises: the Atens, Apollos, and Amors. The term 'near-Earth objects' can include comets as well as asteroids, but we will, mainly, confine ourselves to the latter.

The sources of NEOs are described in Chap. 5 but in summary are:

- Main Belt asteroids
- Centaurs
- Short, intermediate, and long-period comets

Long-period comets may pose a greater threat than asteroids because, although they arrive much less frequently than the latter, they travel at higher speeds and therefore have more impact energy. Their orbits are poorly determined, if known at all, prior to their arrival in the inner Solar System, and therefore there is little chance of mounting a mission to deflect or destroy them.

The number of known NEOs is shown in Fig. 6.1. The graph suggests that most of the larger asteroids, equal to or greater than 1 km in diameter, have been identified but that the number of smaller ones discovered, less than 1 km in diameter, is still growing rapidly.

Potentially Hazardous Asteroids (PHAs)

What is a "potentially hazardous asteroid," and what turns a near-Earth object into one? An NEO becomes a PHA if it makes or may make a close approach to Earth, but how close is close? The NASA's JPL Near-Earth Object Program Office website defines a PHA as one with a:

R. Dymock, *Asteroids and Dwarf Planets and How to Observe Them*,
Astronomers' Observing Guides, DOI 10.1007/978-1-4419-6439-7_6,
© Springer Science+Business Media, LLC 2010

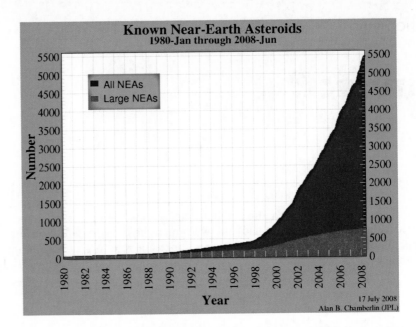

Fig. 6.1. Known near-Earth asteroids. A large NEA has a diameter of approximately 1 km or greater (Credit: NASA/JPL).

Minimum orbit intersection distance (MOID) of 0.05 AU (7.5 million km) or less. The MOID is the minimum distance between the orbits of the PHA and Earth and thus indicates the closest possible approach of the former to the latter.

A diameter of approximately 150 m or greater (this is calculated from the absolute magnitude, which you will learn more about in Chap. 14).

That same website listed 1,012 known PHAs at the end of 2008. The Minor Planet Center and NASA's JPL NEO Program Office maintain various lists of PHAs, close approaches, and potential future Earth impact events.

Craters and Cratering

Earth has been bombarded almost since it came into existence. Our Moon was most likely formed as a result of a collision with a Mars-sized body, and a number of mass extinctions may also be linked to asteroid, or comet, impacts, as explained in Chap. 5. The surfaces of the Moon and Mercury show what can happen to bodies with little or no atmosphere, but Earth's thick atmosphere, continual weathering of its surface, and plate tectonics reduce the number of impacts and quickly eradicate traces of those craters that have formed. Despite this many impact craters have been identified, as shown in Fig. 6.2.

Realization that craters were not necessarily volcanic in origin dawned on the late Gene Shoemaker in 1956 following his investigations of Meteor, or Barringer, crater in Arizona. Figure 6.3 shows a typical weathered crater in Australia that is now named after him.

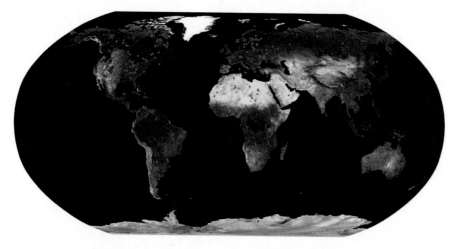

Fig. 6.2. Known impact structures on Earth (Credit: David Kring and Brian Fessler, Lunar and Planetary Institute).

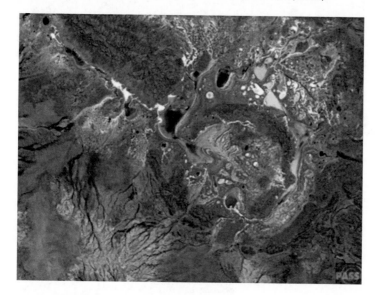

Fig. 6.3. Shoemaker (formerly Teague) crater (Credit: Earth Impact Database, 2006).

What If?

What are the chances of Earth being struck by asteroids, or comets of various sizes, and what would be the effects of such an impact? Figure 6.4 shows the estimated frequency of impact for NEOs of various diameters. The graph shows that, on average, we can expect to be struck by a 1-km diameter asteroid every hundred thousand years or so and by a 'dinosaur killer' 10-km diameter object about every 100,000,000 years. Nature tends not to work to averages, so we could get several such impacts in a single year and then nothing similar for a few million years.

The damage that various size objects could cause is outlined in Table 6.1, based on data provided by Alan Harris.

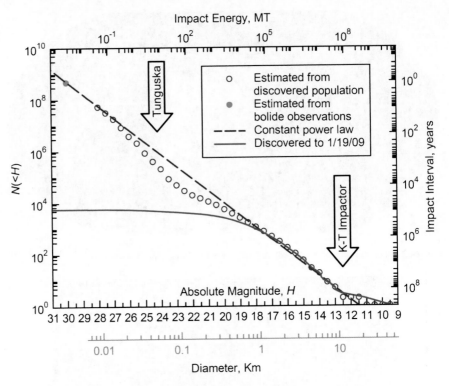

Fig. 6.4. NEO diameter vs. frequency of impact (Credit: Alan Harris, Space Science Institute).

Table 6.1. Effect of airbursts and impacts for asteroids of various sizes

Asteroid diameter	Nature of event	Site of event	
		Ocean	Land
D < 30 m	Airburst in upper atmosphere	No significant damage	
30 m < D < 100 m	Airburst in lower atmosphere	No damage caused	Causes damage similar to nuclear bomb blast above ground, e.g. Tunguska
100 m < D < 1 km	Surface impact	Raises a tsunami that can cause shoreline damage up to a few thousand km distant from the impact point	Makes crater from 2 to 20 km across
1 km < D < 10 km	Surface impact	May raise enough dust into the stratosphere to cause a global climatic catastrophe leading to mass starvation, disease, and general social unrest	
D > 10 km	Surface impact	Possibility of mass extinction, certainly of some species and possibly humans. It is important to distinguish between an event that kills most individuals and a much larger one that causes an extinction. 'Most' is just more than 50%, and an 'extinction' requires nearly all, 99% or more, to be destroyed. Even if there are a few survivors it is highly unlikely that they would be able to find each other to carry on the species. An event that kills off half or more of a species doesn't even show in the geological record, whereas an extinction does	

There is some debate among astronomers as to what size asteroid can reach the ground as opposed to breaking up in the atmosphere. Some say a 50- to 75-m-diameter body would do so whereas others put that diameter at around 200 m (for stony asteroids). A 2007 study by Sandia National Laboratories suggests that smaller

Table 6.2. The Torino Impact Hazard scale

Threat level	Scale	Description of threat
No hazard	0	The likelihood of a collision is zero, or is so low as to be effectively zero
Normal	1	A routine discovery in which a pass near Earth is predicted that poses no unusual level of danger
Meriting attention	2	A discovery of an object making a somewhat close pass near Earth
by astronomers	3	A close encounter, meriting attention by astronomers. Current calculations give a 1% or greater chance of collision capable of localized destruction
	4	A close encounter, meriting attention by astronomers. Current calculations give a 1% or greater chance of collision capable of regional devastation
Threatening	5	A close encounter posing a serious, but still uncertain, threat of regional devastation. Critical attention by astronomers is needed to determine conclusively whether or not a collision will occur
	6	A close encounter by a large object posing a serious but still uncertain threat of a global catastrophe. Critical attention by astronomers is needed to determine conclusively whether or not a collision will occur
	7	A very close encounter by a large object, which if occurring this century, poses an unprecedented but still uncertain threat of a global catastrophe
Certain collisions	8	A collision is certain, capable of causing localized destruction for an impact over land or possibly a tsunami if close offshore. Such events occur on average between once per 50 years and once per several 1,000 years
	9	A collision is certain, capable of causing unprecedented regional devastation for a land impact or the threat of a major tsunami for an ocean impact. Such events occur on average of between once per 10,000 years and once per 100,000 years
	10	A collision is certain, capable of causing global climatic catastrophe that may threaten the future of civilization as we know it, whether impacting land or ocean. Such events occur on average once per 100,000 years, or less often

asteroids, which break up in the atmosphere – airbursts – may carry a bigger punch than originally thought because previous simulations had not taken the forward momentum of the asteroid into account. As this size of asteroid is more numerous in the vicinity of Earth it may be more of a threat than previously supposed.

Two scales are used by astronomers to define the likelihood of a particular asteroid striking Earth and the damage such an impact might cause. The Torino Impact Hazard Scale devised by Richard Binzel, a simplified version of which is shown in Table 6.2, is designed specifically to communicate such hazards to the public, whereas the Palermo Technical Impact Hazard Scale, devised by Steven R. Chesley and colleagues, is used by NEO specialists to assess the risk in more detail.

Discovery

Discovering a new asteroid is not quite as simple as making a few observations and leaving it at that. The discovery process has several parts:

- Initial detection
- Confirmation
- Follow-up observations (usually, but not always, over several orbits, as described in Chap. 2)

Most newly discovered asteroids are/were first imaged by one of five automated surveys operated by professional astronomers: Catalina Sky Survey (CSS), Near-Earth Asteroid Tracking (NEAT), Lincoln Near-Earth Asteroid Research (LINEAR),

Lowell Observatory Near-Earth Object Search (LONEOS), and Spacewatch. LONEOS ceased operations at the end of February 2008 and NEAT in 2007, but the others were still operational in 2009. As far as NEO discoveries are concerned the CSS, with observatories in Australia and the United States, tops the pile. This may all change when the Panoramic Survey Telescope and Rapid Response System (Pan-STARRS) and Large Synoptic Survey Telescope (LSST) become operational. These observatories will be able to detect much fainter objects and image far greater areas of sky in a given time than the existing surveys.

A typical search program involves taking several images of a particular area of sky over the course of an hour. These images will then be analyzed by moving object detection software to eliminate stationary objects (e.g., stars), blemishes, and cosmic ray hits. If a moving object is detected its position will be compared with known asteroids and comets and a report submitted to the Minor Planet Center (MPC).

The MPC publishes details of newly found objects on its NEO Confirmation Page (NEOCP), allowing other observers, professional and amateur, to attempt to image the object and thus confirm the discovery. This is not as simple as it may seem as, at this stage, the initial orbit computed by the MPC is not well defined and thus predictions of future positions may be somewhat uncertain. However the MPC publishes maps of such uncertainties, allowing observers to search a specific area of sky for the new object. Should you wish to calculate the orbit and ephemeris there are several pieces of software that will enable you to do so, which are listed in Appendix B of this book, Resources.

As an example, a fast-moving object (FMO) was discovered by LINEAR on July 11, 2005, to which they assigned the designation AU52949. The data published on the NEOCP enabled UK amateur astronomer Peter Birtwhistle to search for and find this FMO. Figure 6.5 shows the possible positions of the asteroid overlaid with Birtwhistle's CCD field of view. As can be seen many images had to be obtained before the object was finally located on the bottom two images. The initial discovery and subsequent follow up observations enabled the object to be given the provisional designation 2005 NG$_{56}$. Discoveries are announced by the MPC via Minor Planet Electronic Circulars (MPECs). The professionals will probably find most NEOs, threatening or not, but the role of the amateur in discovery confirmation should not be underestimated, and amateurs do also still make discoveries.

Uncertainty Becoming Certainty (Hit or Miss)

As can be construed from Fig. 6.5 the orbit initially calculated is somewhat uncertain, but why is this? Imagine that you are watching a golf tournament but are only able to see the path taken by a ball during a very short period of its flight. Knowing, or assuming, the tee from which the ball has been hit and thus the green aimed for you can make a very rough estimate of where it will land. Of course the ball will spin and swerve during its flight, making that estimate even more difficult.

Similarly, when the orbit of an asteroid is calculated from a very few observations it is difficult to be precise. One can make a few assumptions as to its inclination, eccentricity, and semi-major axis, and that the object is orbiting the Sun. What is

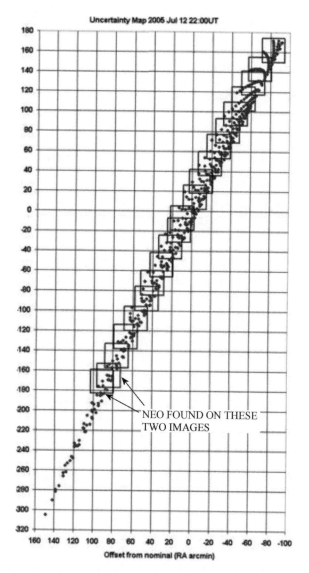

Fig. 6.5. Uncertainty map overlaid with search pattern (Credit: Minor Planet Center/Peter Birtwhistle, Great Shefford Observatory, UK).

needed are more observations over a longer period of time – ideally one or more complete orbits – and professional astronomers find it difficult to obtain the telescope time to do this. This is where amateurs play a very important role, as continual monitoring of the positions of asteroids with reasonably well-defined orbits is a somewhat easier task for those with more modest equipment. How this can be achieved is described in Chap. 10.

New observations are not the only way of confirming discoveries and establishing orbits. The NASA SkyMorph facility allows astronomers to search through many years' worth of accumulated imagery. Such prediscovery, or pre-covery, observations

Impact?

may significantly increase the period of observations, or observational arc, of a particular asteroid, allowing its orbit to be determined with much greater accuracy.

Tracking NEOs is not the sole prerogative of the optical observatories. Radar observations of potential impactors are useful, if not essential, as they can tell us much about those asteroids such as size, shape, spin, structure, and surface properties. Of great importance, they can also determine their positions and therefore their orbits with great precision. Simultaneous imaging by amateur astronomers can also achieve accurate and timely calculation of positions of NEOs as is explained in Chap. 11.

Near-Earth asteroids are monitored and orbits and impact risks calculated by the NASA/JPL Sentry and the University of Pisa's (Near Earth Object Dynamics Site) CLOMON2 systems. The Sentry system uses all observations, optical and radar, accepted by the MPC to continually update asteroid orbits. Close approaches to Earth for up to 100 years are calculated for many thousands of "Virtual Asteroids," the orbits of which fit the observations reasonably well. If, for example, the paths of 100,000 Virtual Asteroids are calculated and one of these were predicted to strike Earth, then the probability of impact for that asteroid at that time would be 1/100,000. That particular Virtual Asteroid is then designated as a Virtual Impactor. Figure 6.6 shows a typical Uncertainty Region, which includes the paths of all the Virtual Asteroids. In case 1 none of these will strike Earth, but in case 2 some of them will since the distance, D, from the center of Earth to the center of the Uncertainty Region is less than one Earth radius, and these are therefore Virtual Impactors.

Asteroid (99942) Apophis has been much in the news as being a possible impactor. The chances of this happening are very low at present – about 1 in 43,000 between 2036 and 2069, which translates as a 99.9977% chance that it will miss Earth.

The CLOMON2 monitoring system is run by the University of Pisa, Italy. As does Sentry it utilizes observations collected by the MPC. CLOMON2 is updated on a daily basis, and the observations used are listed on the NEODyS website by object and by observatory. The Risk Page on that website lists all asteroids that pose a threat sometime between the present day and 2080 (soon to be extended to 2090). Amateur astronomers can use this as a resource to track down those asteroids that

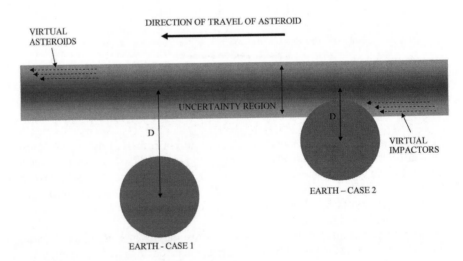

Fig. 6.6. Virtual Asteroids and Virtual Impactors (Diagram by the author).

pose a threat. Some of the asteroids on the 'Lost Objects' list have not been observed for some time – 8 years not being untypical. Finding these objects would be an interesting and extremely useful project.

Who Watches the Watchers?

A good question but, at this time, one without an answer. One might hope, indeed expect, that there would be some organization at the national or international level that would review the danger of a possible impact and prescribe action to be taken to reduce or eliminate the threat and protect the population likely to be affected. Unfortunately this is not the case. On the NASA/JPL NEO Program Sentry Risk Table webpage you will find the statement 'Whenever a potential impact is detected it will be analyzed and the results immediately published here, except in unusual cases where an IAU Technical Review is underway.'

So what is an IAU (International Astronomical Union) Technical Review? The IAU position is that 'This review procedure is encouraged for any prediction that is at a level equal to or greater than zero on the Palermo Technical Scale. In most cases, such events will fall at a value of 1 or higher on the 0–10 point Torino Scale.' This Technical Review process, more recently absorbed into the Advisory Committee on Hazards of Near Earth-Objects, reviewed several cases around the year 2000 but since that time nothing much seems to have happened.

None of the above is mandatory and, as mentioned earlier, there is software available that allows anyone to calculate an orbit from observational data. Therefore it is possible that the first announcement of a possible impact could come from an astronomer, amateur or professional, rather than via Sentry, CLOMON2, or the International Astronomical Union. This leaves lots of room for confusion!

Several conferences have been held to propose solutions and organizations such as Spaceguard. The B612 Foundation and the Association of Space Explorers have pressed and are pressing for this final part of the jigsaw to be put in place. In September 2008 the latter submitted a report 'Asteroid threats: a call for global response' to the United Nations 'for consideration and subsequent action by the United Nations; its goal is to assist the international community in preventing loss of life and property resulting from an asteroid impact on Earth.' We all await a response....

Of course there may not be time for too much debate as, for example, asteroid 2008 TC_3 was discovered on October 6, 2008, only 19 h before impact. Admittedly this was a small object that exploded in the atmosphere and did not, as far as is known, cause any damage. Other NEOs have been detected just before (both by LINEAR): 2004 FH in 2004, or only after, 2002 MN in 2002, their closest approach.

Deflect or Destroy?

What is our best chance of protecting ourselves? The report mentioned above dealt mainly with how an asteroid might be deflected away from Earth – threat mitigation is the term most often used in this context.

Deflection

An asteroid passing close to Earth may pass though one or more small regions that guarantee an impact sometime in the future. These regions are known as keyholes. For example asteroid (99942) Apophis will pass close to Earth in 2029, and, if it passes through a keyhole just 610 m wide, it will receive a gravitational nudge that will put it on a collision course with our planet in 2036. The vast majority of potential impactors will have passed through a keyhole in the previous years or tens of years.

For a deflection to be successful the orbit of an asteroid must be changed such that not only does it miss Earth at the predicted time of impact but that such changes do not direct it through a keyhole, thus leading to a future impact. If a potential impactor could be identified, say, 10 years in advance, then to achieve the required deflection a change in its velocity of only a few centimeters per second is all that is required. This is not a lot considering that a typical NEO might be moving at 3 million cm/s.

Many ideas have been floated as to how to deflect an asteroid including:

- A number of spacecraft focusing sunlight onto the surface of an asteroid, the resulting vaporization of the surface causing a small thrust in the opposite direction. Best suited to smaller asteroids, a similar result might also be achieved using a laser.
- The gravity tractor method by which a spacecraft flies alongside an asteroid and uses the tiny gravitational force it exerts to nudge the asteroid into a new orbit.
- Exploding a nuclear weapon, or a series of them, near the asteroid – but this might cause the asteroid to break up, and we would then have to cope with many small, possibly radioactive, asteroids instead of one large one.
- Ramming the asteroid with a spacecraft, as successfully demonstrated when the Impactor part of the *Deep Impact* spacecraft navigated itself to a collision with comet Tempel 1 on July 4, 2005. The European Space Agency (ESA) has proposed *Don Quixote* mission consists of an orbiter spacecraft, *Sancho*, and an impactor, *Hidalgo* (see Fig. 6.7).

Fig. 6.7. Hidalgo impactor – part of ESA's Don Quixote mission (Credit: ESA-AOES Medialab).

- Utilizing a mass driver whereby the asteroid surface is dug up and the material ejected to push the asteroid in the opposite direction.
- Attaching a thruster to the surface of the asteroid.
- Taking advantage of the Yarkovsky effect (explained in Chap. 5) by coating part of the surface of the asteroid with a white substance, thus affecting the amount of heat absorbed and reradiated.
- Attaching a solar sail to the asteroid and allowing solar radiation pressure to come into play.

Destruction

This is not the favored option, as exploding a nuclear weapon on or under the surface of an asteroid may result in a swarm of small, possibly contaminated, objects hitting Earth rather than one large one. As was mentioned in Chap. 2, many asteroids are rubble piles rather than solid bodies, against which a nuclear explosion would be even more likely to create a swarm rather than lead to a deflection. The rotation rate of an asteroid is an indicator of its composition – a fast rotator is unlikely to be a rubble pile. As will be explained in Chap. 13, determination of rotational periods by constructing lightcurves is a suitable, and not too difficult, project for the amateur astronomer.

That concludes the descriptive section of the book. Now we move on to the more practical matters of observing and imaging asteroids and dwarf planets.

Part II

Observing Guide

Chapter 7

Observatories

Now that we have some understanding of what asteroids are and where they are located in the Solar System, it is time to find out how to observe them. Before getting into detail of visual observing and imaging, a tour of the equipment and observatories used and owned by amateur and professional astronomers will help to set the scene.

Degrees of automation vary from purely manual operation to remote and/or automatic control of telescope, camera, filters, and dome or roof. Software is available off-the-shelf to support these functions. Other images of telescopes and accessories can be found in the succeeding chapters and some relevant books are mentioned in Appendix B of this book.

A quick word on telescope mounts, which come in two varieties: equatorial and altitude-azimuth (alt-az). For all the activities described in the observing section of this book the former are preferable.

Temporary or Portable Set-Up

Although not always the case, this mode of operation is generally more suitable for visual observing (Chap. 8), unencumbered by electronic devices and their necessary cabling, than it is for imaging. When the time taken to prepare to observe, and to disassemble your equipment, becomes a significant proportion of the time available to you, then it rather takes the edge off things. It may be that unless the chances of a clear sky are 100% you will be put off observing altogether at that particular time. Of course if travel is necessary, to put yourself on the track of an occultation (Chap. 15), for example, then so be it, but do give yourself extra time on arrival at your chosen site to ensure all is in perfect working order.

The minimum requirement is a stable and dry surface on which to site your telescope. If this is on your own property, then the position of your mount can be marked so that orientation and polar alignment is simplified. Figure 7.1 shows British Astronomical Association (BAA) member Martin Mobberley's 250 mm (10 in.) Orion UK SPX reflector, which can be rolled out of its 'kennel' along the strips of flexible plastic into a pre-set position and be ready for observing within minutes. The smoothness of the plastic strips and the underlying surface help to prevent any unnecessary jolting of the telescope during its travels. Mobberley has been observing and imaging a wide range of celestial objects for 40 years and, in addition, has written a number of books, including: *The New Amateur Astronomer* and *Lunar and Planetary Webcam User's Guide*, both published by Springer.

R. Dymock, *Asteroids and Dwarf Planets and How to Observe Them*,
Astronomers' Observing Guides, DOI 10.1007/978-1-4419-6439-7_7,
© Springer Science+Business Media, LLC 2010

Fig. 7.1. Martin Mobberley with telescope pulled out of its 'kennel' and ready to use (Credit: Martin Mobberley).

Backyard Observatories

Complete observatories can be purchased, but many amateurs do build their own. The range is quite varied: domes, roll-off roofs, roll-away structures, and mini-observatories just large enough to house a telescope, an imager, and the associated electronics.

Observatory Domes

When you think 'observatory' you often think 'dome.' Circular structures with rotating domes on top can be seen in amateurs' back gardens, local astronomical societies' sites, and on the bleak mountaintop observatories of professional astronomers. Peter Birtwhistle's dome, in the not-so-bleak setting of his garden in Great Shefford, England, is shown in Fig. 7.2 (MPC station code J95 – see Chap. 10 for how to obtain such a code). The Sky Domes 2.4-m (8-ft) diameter fiberglass dome houses an 0.4-m (16-in.) Meade LX200 GPS telescope. Birtwhistle is one of the world's foremost amateur asteroid observers, who by the beginning of January 2009 had amassed a total of 10,000 positions of near-Earth objects. Examples of his work are described in Chaps. 6 and 11.

Roll-Off Roof Observatories

Figure 7.3 shows one such example built by UK amateur Michael Clarke in 2006. The 8 × 8 ft roll-off roof observatory, named 'Gargoyle,' houses an 80-mm f/7 Triplet refractor mounted on a Vixen GP equatorial mount located on a pyramidal pier.

Fig. 7.2. Summer arrives at Peter Birtwhistle's Great Shefford Observatory (Credit: Molly Birtwhistle, Peter's daughter).

Fig. 7.3. Michael Clarke's roll-off roof observatory (Credit: Michael Clarke).

Examples of Mike's work with a Canon 350D DSLR camera mounted on the telescope are shown in Chap. 9.

At the other end of the scale, Irish amateur Eamonn Ansbro uses an 0.9-m (36-in.) reflector to search for Edgeworth–Kuiper Belt objects (Chap. 11). This large, by amateur standards, telescope is located at Ansbro's Kingsland Observatory

Fig. 7.4. Eamonn Ansbro's observatory (Credit Eamonn Ansbro).

(MPC station code J62) in County Roscommon, Ireland (see Fig. 7.4). One advantage of a roll-off roof is that, unlike a dome, it does not have to be rotated as the telescope follows an object. Rolling off part of the sidewall allows the telescope to be aligned on objects at a lower altitude than if the roof alone is retracted.

Roll-Away Shed

You can roll away the roof, but you can also roll away the complete protecting structure! UK amateur Nick James constructed his observatory (MPC station code 970) using a standard 1.8 × 1.5-m (6 × 5 ft) garden shed running on four V-groove pulley wheels (two on each side) along T-section angle-iron pieces about 6-m (20-ft) long. Figure 7.5 shows the shed rolled away with the telescope, an 11-in. SCT equipped with an SBIG ST9 CCD, in its operating position. To ensure stability the telescope is mounted on an 8-inch-diameter steel plinth embedded in about 1 ton of concrete. The telescope and CCD camera are operated via a PC situated next to the telescope, which is connected to a PC in Nick's study. James's astronomical interests are varied and include imaging near-Earth asteroids and comets.

A Remote-Controlled Observatory

If your space for a backyard observatory is limited and your preference is to image from the comfort of your house then the BAA's Asteroid and Remote Planets Section Director, Richard Miles,' observatory (MPC station code, J77), Fig. 7.6, may be the type of thing you are looking for. The upper section, which hinges back,

Fig. 7.5. Nick James's roll-off shed observatory ready for use with pillar extended (Credit: Nick James).

Fig. 7.6. Richard Miles' observatory (Credit: R. Miles).

is just 102-cm wide × 81-cm deep × 73-cm high, and the overall height of this wooden structure is 153 cm. Each of the three telescopes, a Celestron 28-cm f/10 Schmidt–Cassegrain and two Takahashi 6-cm f/5.9 refractors, all on the same German equatorial mount, is controlled by a laptop computer situated in the 'box,' and the system is run remotely from Richard's study. An example of his work can be found in Appendix C of this book.

Amateur Astronomical Groups

Membership in a local astronomical group, in particular groups with their own permanent observatories, can be advantageous. Not only will you enjoy the company and be able to share experiences with like-minded people, but you will also, most likely, have access to some quite powerful telescopes and maybe some with interesting histories. Open evenings allow you to explain the many varied aspects of the night sky to members of the public (though to the uninitiated asteroids are not very exciting). If you are able to visually observe or image Pluto then a lively discussion on its demotion from planet to dwarf planet may result. However, one does have to learn to deal with a totally unimpressed 'Mr. Grumpy' who complains, at 1:00 AM, that Venus looks like a 'broken fingernail.' And what do you say to the person who insists that he observes with a 'gamma-ray telescope'?

The Hampshire Astronomical Group's (HAG) site, located in Clanfield, England, is home to several telescopes, each housed in its own dome. Figure 7.7 shows the 0.61-m (24-in.) reflector on an English yoke mount. This telescope is used for both visual observing and imaging and is controlled from a console out of view to the right of the image. Also used to make astrometric measurements of asteroids, this telescope has the MPC station code J84. With the exception of the main mirror and eyepieces the telescope, and dome, were constructed by members of the group.

A dome once located at what was the Royal Greenwich Observatory at Herstmonceux, England, houses a 127-mm (5-in.) Cooke refractor built around 1890, and a 114-mm (4.5-in.) Beck, Beck, and Smith refractor built in the early 1880s (see Fig. 7.8). The Cooke refractor is believed to have been previously owned by Sir Harold Spencer Jones, the tenth Astronomer Royal of England from 1933 to 1955. Having observed asteroids (4) Vesta and (8) Flora from my own back garden I observed my third asteroid, (89) Julia, through the Cooke refractor on November 12, 1997.

Fig. 7.7. HAG's 0.6-m (24-in.) reflector (Credit: Hampshire Astronomical Group).

Fig. 7.8. HAG's recently refurbished twin-mounted 114-mm (4.5-in.) Beck (bottom) and 127-mm (5-in.) Cooke (top) refractors (Credit: Hampshire Astronomical Group).

Robotic Telescopes

A significant advance in amateur astronomy in recent years has been the setting up and use of remotely operated or robotic telescopes. Access to telescopes positioned at various longitudes and in both hemispheres means that you can track an object when it has moved out of your field of view and observing does not necessarily have to be a night time occupation! Use of such facilities need not be expensive, especially when compared with the cost of setting up one's own observatory from scratch. Robotic telescopes are available to suit the beginner – SLOOH, for example – and the more experienced observer, e.g., Sierra Stars Observatory Network (SSON), Skylive, Bradford Robotic Telescope (BRT), and Global Rent-A-Scope (GRAS).

There is much to be said for the satisfaction of observing through the eyepiece and most, if not all, amateurs start out this way. However, this avenue is not open to all. There may be no suitable place to site a telescope of their own, no convenient local astronomical society, or a disability that limits their movement. So not only do robotic facilities offer an alternative to amateurs without access to a telescope, they also allow those with fewer resources to enjoy this hobby.

Typical of the automated robotic telescopes available are those of the Sierra Stars Observatory Network (SSON). Two telescopes are available to amateurs, professionals, and schools: an 0.61-m (24-in.) Optical Mechanics Nighthawk CC06 Cassegrain with a Finger Lakes ProLine CCD camera and photometric filters, located in Alpine County, California (MPC station code, G68) (see Fig. 7.9), and a 0.37-m (14.5-in.) Optical Mechanics Cassegrain telescope, the same CCD camera as the 0.61 m telescope but with color filters, at Sonita, Arizona, owned by the University of Iowa (MPC station code, 857).

A recent interesting discovery, made using the larger telescope, was that of a Jupiter L4 Trojan asteroid (see Chap. 3 for more on asteroid groups), subsequently given the designation 2009 UZ_{18}, by Bill Dillon and Don Wells. The author used this telescope,

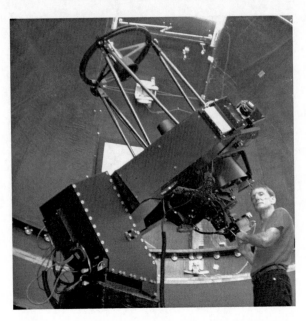

Fig. 7.9. SSON's 0.61-m (24-in.) Optical Mechanics, Inc. Nighthawk CC06 Cassegrain telescope and CEO Rich Williams (Credit: Rich Williams).

needing the photometric V filter, to image asteroid (01909) Alekhin to obtain a phase curve (see Chap. 14) and a lightcurve (see Chap. 12) for this asteroid. One big advantage of SSON is that there is no need to be at one's PC when the images are being taken. You select an asteroid from a list (or enter coordinates) and enter time, date, duration, number of, and time between exposures. The images are scheduled and taken completely automatically and are available for download the next day.

Professional Observatories

Catalina Sky Survey (CSS)

Other professional surveys, including LINEAR, Spacewatch, NEAT, and LONEOS, may have discovered more asteroids in total, but CSS holds the record for NEO discoveries. This survey is a consortium of three cooperating groups: the original Catalina Sky Survey on Mt. Bigelow in Arizona, after which the project is named, the Mt. Lemon Survey (MLSS) at the summit of Mt Lemmon also in Arizona, and the Siding Spring Survey (SSS), New South Wales, Australia.

The objectives of this survey are to discover and perform follow-up observations of NEOs and assess the threat posed by PHAs, by determining their size, density, shape, albedo, and velocity (see Chap. 6 for more on NEOs and PHAs and Chap. 14 for a description of albedo).

The SSS (MPC station code, E12) (see Fig. 7.10), uses an 0.5-m (20-in.) Schmidt telescope and a Spectral Instruments, Inc. CCD camera identical to that attached to the Mt. Bigelow telescope. This telescope is primarily used by CSS to perform follow-up observations of NEOs, but six NEOs, one PHA, and one comet have been discovered with it.

Fig. 7.10. Siding Spring Survey observatory and telescope (Credit: Siding Spring Survey).

Between 1990 and 1996 the Anglo-Australian Near Earth Asteroid Survey (AANEAS), later renamed Spaceguard Australia, made use of the photographic plates obtained by the UK's Schmidt Telescope located at the Siding Spring Observatory. Typical exposure times for these plates were 60–180 min, and asteroids could be identified by the short trails they left on the plates. AANEAS also used the 1-m (40-in.) Siding Spring Observatory telescope and occasionally the 3.9-m (156-in.) Anglo-Australian Telescope for follow-up astrometry.

Panoramic Survey Telescope and Rapid Response System (Pan-STARRS)

Pan-STARRS is being developed at the University of Hawaii's Institute for Astronomy. Its location is yet to be confirmed, but Mauna Kea on Hawaii is the preferred option. Its objective is to discover and characterize Earth-approaching objects, both asteroids and comets, that might pose a danger to our planet.

The completed system will comprise four individual optical trains, each with a 1.8-m (72-in.) mirror and a 1.4 billion pixel CCD camera, which will observe the same region of sky simultaneously. The whole available sky will be observed three times each night during the dark period in each lunar cycle; objects as faint as magnitude 24 should be within its grasp.

A single-mirror prototype, PS1 (see Fig. 7.11), was completed and a CCD camera installed by August 2007. Commissioning will be followed by a 3.5-year-long series of tests and studies.

Space Missions

If you can't bring asteroids or dwarf planets to you by means of collecting the reflected sunlight in your Mark 1 eyeball or any of a variety of imagers, then the alternative is to go to those bodies. Of course this opportunity is only open to

Fig. 7.11. Pan-STARRS PS1 prototype on Haleakala, Hawaii (Credit: Brett Simison).

professional astronomers, but amateurs can share in those journeys and obtain images and data via the internet.

Current operational missions (2009) include:

- *Dawn*, on its way to the dwarf planet (1) Ceres and asteroid (4) Vesta (see Fig. 7.12)
- *Hayabusa*, returning home after collecting a sample from asteroid 1998 SF_{36} Itokawa
- *New Horizons*, scheduled to arrive at dwarf planet (134340) Pluto in 2015

In addition to these:

- The Near-Earth Object Surveillance SATellite (NEOSSAT) is being designed to discover near-Earth asteroids (and track satellites).
- *Don Quixote* will test a method of deflecting an Earth-approaching asteroid.
- A sample of the surface material of asteroid 101955 will be returned to Earth by the Origins Spectral Interpretation Resource Identification and Security (OSIRIS) spacecraft.
- The MErcury Surface Space Environment GEochemistry and Ranging (MESSENGER) mission will look for the elusive Vulcanoids as a secondary objective to its primary target of the planet Mercury.
- The Near-Earth Asteroid Rendezvous (NEAR) spacecraft completed its trip to asteroid (433) Eros with a soft landing on that asteroid in 2001.

Fig. 7.12. Artist's rendition of the *Dawn* spacecraft leaving Earth for (1) Ceres and (4) Vesta (Credit: NASA/JPL).

Other Earth-orbiting observatories have targeted or will observe asteroids in addition to many other objects. These include:

- The HIgh Precision PARallax Collecting Satellite (HIPPARCOS)
- The Wide-field Infrared Survey Explorer (WISE) mission
- The InfraRed Astronomy Satellite (IRAS)

As you will have seen here, and elsewhere in this book, telescopes, imagers, accessories, and observing sites are many and varied. The following chapters describe how their various combinations can be used to observe and image asteroids.

Chapter 8

Visual Observing

From reading the previous chapter you now have some idea of what is involved in chasing down these 'vermin of the skies,' as they were once referred to because their tracks spoiled photographic plates. Asteroids are not alone in being called that, as that description has also been given to artificial satellites and flocks of birds.

Finding asteroids by observing them through the eyepiece of a telescope or, for a few brighter ones, using binoculars is reasonably easy and is a good introduction to this sphere of astronomy and, in general, finding your way around the night sky. As mentioned in the opening chapter it is assumed that the reader is computer literate, reasonably familiar with the night sky, and understands how to set up and use a telescope. We are making somewhat of an exception to that rule in this chapter, as it may appeal more to the less experienced among you, who might appreciate a little extra guidance.

What kind and size of telescope (very briefly)? A 6 in (15 cm) reflector is a good starting point, but you may want something larger if you are going to get really serious – a 10 in. (25 cm), 12 in. (30 cm), or 14 in. (36 cm) Newtonian reflector or Schmidt–Cassegrain, for example. If you have an astronomical society close at hand you can 'try before you buy' or, if such a group has a permanent observatory, you may decide that using their telescopes is more your style – 'try and not buy' you might call it. Nearly all modern off-the-shelf telescopes have a drive motor, at least in Right Ascension, and this is pretty much a necessity. It is very hard to find your way around the sky with an undriven telescope mount. Every time you refer to your star chart and return to the eyepiece you will see a different star field because Earth has moved and your telescope has not!

You should be able to see a number of the brighter asteroids using, typically, 10×50 binoculars. If, for example, your binoculars will allow you to see objects as faint as magnitude 8 or 9, then you may be able to find five or six asteroids in any year. Because of this small number of targets typically available binoculars are not recommended for asteroid observation, and their use is not covered in this chapter. However, if you are combining asteroid observing with other aspects of astronomy such as comet observation, then why not give it a try? A reclining chair is a great help, but be careful not to fall asleep! There are a number of commercial binocular supports available, and, of course, image-stabilized binoculars are a great help but somewhat more expensive than the regular kind.

Unless you live at or have access to a truly dark site and have excellent vision, it is extremely unlikely that you will be able to see any asteroids with the naked eye. However, attempting to do so would make an interesting project – more about this later in this chapter.

R. Dymock, *Asteroids and Dwarf Planets and How to Observe Them*,
Astronomers' Observing Guides, DOI 10.1007/978-1-4419-6439-7_8,
© Springer Science+Business Media, LLC 2010

Visual
Observing

Getting Started

One complication of asteroid observing is that their positions are ever changing with respect to the background stars. Most, if not all, printed star atlases don't show stars to a faint enough magnitude, and none will show the positions of asteroids. The most detailed charts are, of course, also the most expensive and therefore not the sort of thing you want to subject to the cold and damp outdoors (although the best are laminated and even more expensive). Astronomy magazines often publish charts for the very brightest of asteroids, but this will not give you a very large sample to work with. To be able to find and visually observe a good crop of asteroids you really need charts, and of course a telescope, that go down to magnitude 11 or 12.

By far the best method is to invest in a laptop PC that you can take with you to wherever you happen to be observing. A cheap or second hand one will do, and it really does make life a lot easier. You will also need planetarium software that has the facility to download the latest asteroid orbital elements available from either the Lowell Observatory or the Minor Planet Center. This author mostly uses *Megastar* so the examples in this chapter are produced using that software. *Guide* is another popular package, and the Minor Planet Observer's *Asteroid Viewing Guide* provides data and charts for asteroids observable from a specific location at a selected date and time.

You could of course print out your charts using your PC with the asteroids plotted, but, having done so, it will then cloud over and you have destroyed another tree. A run of bad weather soon turns those trees into forests! Printing charts from your PC is still a better bet than investing in an expensive star atlas. If you drop and tread on a few sheets of computer printout while observing or leave them at your local astronomical group it is no great loss, but if you were to do the same with your priceless star atlas – aaargh!

Checklists can be very useful, and, as your skills grow and your equipment becomes ever more complex, you may find they help even more. So, to help you through the night and generally make your observing life easier, here is one for starters:

- Be warm and safe. You may end up looking like an arctic explorer, but if you are warm you will stay outside longer and get more enjoyment out of your observing. **BE SAFE!!!** It is not unheard of for an amateur astronomer to trip over in the dark and injure sustain an injury.
- Ensure your telescope is sited on a stable and dry base. I started with paving stones laid on soil but of course they wobble (and so does your telescope) as you move around. Ideally the telescope should be permanently sited or at least kept in the environment in which it is to be used. If you progress to using electronic cameras and their associated electronics it is best to leave all this stuff permanently connected. It saves set-up time and problems due to poor, damaged, or wrongly plugged connections.
- Polar alignment is necessary if you want your telescope to track well. Alignment differs for different makes of telescope, but the end result will be that the polar axis is aligned with the pole star. (Polaris is not exactly aligned with the North Celestial Pole, but it is close enough for starters.) To very roughly align your telescope point the polar or right ascension axis due north and elevate the tube to the local latitude. The advantage of having a permanent site is that you don't have to polar align every time you want to observe.

- To help you star-hop around the sky ensure your finderscope is aligned with your telescope tube so that the object centered in it is also centered in the eyepiece field of view (FOV). This is done by centering a bright star (such as Aldebaran in the constellation of Taurus) in the eyepiece FOV and adjusting the finderscope so that the same star is at the center of its FOV. If the alignment is way off start by sighting on the star along the edge of your telescope tube.
- If you have a telescope with a 'go-to' facility you will need to align your telescope on a number of bright stars, three is best, so that the telescope's controller knows where it is. This procedure may vary from 'scope to 'scope, so you should refer to its handbook to do this.

One of the trickier aspects of visual observing is understanding what you are seeing through the finderscope or eyepiece, i.e., the orientation and size of your field of view. (Once using a different telescope I hadn't realized that the finder FOV was not the same orientation as my own and spent most of the evening 'lost'.) This is best done by using an area of the sky that has some very obvious features, e.g., a star cluster such as the Hyades in Taurus. (Aldebaran is close by and could be used to centre your finder as mentioned above.)

If you are using a computer and planetarium software such as *Megastar* you can overlay circles relating to your finder and various eyepieces. This really does make finding your way around the sky much easier. If you are using printed charts then make up some cardboard overlays equivalent to the sizes of the various FOVs you are likely to encounter or use a flowchart template. Planetarium software will also allow you to orientate the star chart to match your FOV (north or south up, mirror, for example). A further advantage of computerized star charts is that you can adjust the magnitude of the stars displayed to match what you see through your finder or eyepiece. Figure 8.1, the Hyades cluster south up, shows stars to magnitude

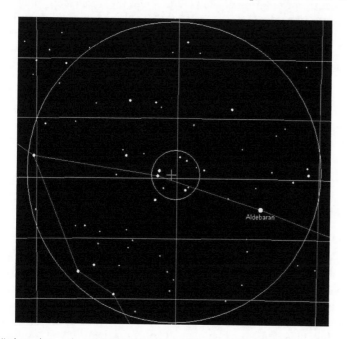

Fig. 8.1. Hyades star cluster, south up with finder, 5° FOV, and eyepiece, 50 arcmin FOV, circles overlaid. Stars to magnitude 8 shown (Credit: *Megastar*, ELB Software).

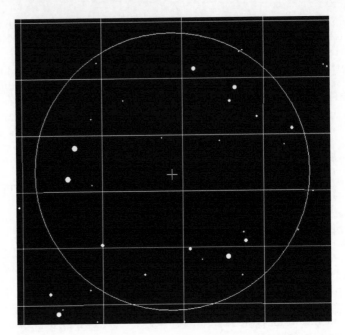

Fig. 8.2. Hyades star cluster, south up, with eyepiece FOV, 50 arcsec, overlaid. Stars to magnitude 12 shown (Credit: *Megastar*, ELB Software).

8 – typical for a finderscope; overlays for a 5° FOV finder; and a 48 arcmin FOV eyepiece. Figure 8.2 shows that same field but as it would look through the eyepiece. Be aware that the orientation of the FOV through the finderscope may be different to that through the eyepiece. An FOV of somewhere between 30 and 60 arcmin should include a reasonable number of comparison stars, but do experiment with different focal length/magnification eyepieces for best results, e.g., not too many or too few stars, good visibility of asteroid and field stars, easy to focus, and minimum vibration.

Most, if not all, star charts, computerized or in paper form, show brighter stars as larger and larger dots. This can be misleading, as it tends to make the stars look closer together than they do through the eyepiece. If your planetarium program has a facility to plot all stars the same size try it out to see if identification of the star field is made easier.

Limiting Magnitude

Magnitude has been mentioned several times, and it is very useful to know how faint you can go with your set-up, i.e., what is its limiting magnitude, and that isn't an easy question to answer. Knowing your limiting magnitude will also enable you to adjust your planetarium software accordingly. Matching what you see to a star chart can be quite difficult if you see more or fewer stars than are shown on the chart. There are formulae for calculating what you might see through various telescopes or binoculars, but so much depends on your own 'optical train' and the sky conditions at your observing site that it is hard to be too specific.

A simple equation that calculates the gain in magnitude using a telescope over what can be seen with the naked eye is:

$$G = 5 \times \mathrm{Log}(D/P)$$

where G = gain in magnitude, D = diameter of the telescope's main mirror or primary lens, and P = diameter of the observer's dark-adapted pupil. So if D = 250 mm (10 in.) and P = 6 mm (0.25 in.) then G = 8. If your naked-eye limiting magnitude is 4 you should be able to see stars as faint as magnitude 12 with such a telescope – typical for a location on the edge of an urban area. A pair of 10 × 50 binoculars will generally enable you to see stars to magnitude 8.

Note that as you dip into the murk along the horizon you can easily lose a magnitude or two. For a more pristine sky you may be able to go fainter (and performance may vary considerably during the night or from one night to the next).

Finding the Target

There are a number of ways of finding your way around the night sky to whatever it is you want to observe, including:

- Simply moving, or 'star-hopping,' from a bright star close to your target using a low power eyepiece. If you want to make absolutely sure of where you are and what you are seeing you can overlay multiple finderscope and eyepiece circles on your star charts to act as stepping stones from your starting point to your target. But move along one axis at a time; then, if you don't recognize the star field, you can easily backtrack to your starting point.
- Using setting circles – scales on the telescope fixed to the Right Ascension (RA) and Declination (Dec) axis. To use these the telescope needs to be polar aligned, and then pointed at a bright star, and the setting circles set to that star's RA and Dec.
- Using a 'go-to' facility to move to a particular set of co-ordinates. This can be built into the telescope controller or operated via a link from a PC to the controller.

A good method is to plot the track of the asteroid using *Megastar* and then use its 'go-to' telescope control feature to point the telescope. Do ensure that your asteroid data (orbital elements) are up-to-date by downloading the latest orbital elements – usually from the Lowell Observatory or the Minor Planet Center, but refer to your star charting software as to how to do this.

Targets for Tonight

Main Belt asteroids are your likely targets; they are the brighter ones and move at a pace that is detectable in an hour or two, which will help you to confirm your identification. Near-Earth objects and more distant objects are usually too faint for visual observation. The easiest way to identify the area of sky most suitable for observing is to use a planisphere and locate the ecliptic, as most of your targets will be in that area – Main Belt asteroids generally having low inclinations.

A few other tips for ease of observing: to avoid the low-altitude murk try and select asteroids that are at least 20–30° above the horizon; choose asteroids that are in the southeast, as this will give you more time to observe them before they dip down towards the western horizon. Time can easily catch you out. Most sources quote time as Universal Time (UT), so you need to factor in your time zone and correct for Summer Time, Daylight Saving Time, or its equivalent where you live. As an example let's look at the sky from northern mid-latitudes on the evening of March 1, 2009. You can use your star charting software to plot the visible asteroids, but alternatives are:

- The CalSky website. An excellent facility which allows you to select asteroids, plot charts and much more. Very well worth a visit.
- The 'What's Observable' website run by NASA JPL's Solar System Division. Here you can enter date, time, location, and observer constraints (limiting magnitude, for example) and produce a listing of potential targets. Table 8.1 shows a selection of asteroids and data from one such query.
- The Astronomical League's 'What's Up Doc' website.
- The *Handbook of the British Astronomical Association* and the website of its Asteroids and Remote Planets Section.
- The Heavens-Above website.
- The Minor Planet Center website. You will need to obtain an ephemeris for your location and then plot the positions obtained on a star chart.

Scanning this list suggests that (27) Euterpe would be a good target – reasonably bright, in the southeast, and approximately 50° in altitude. Figure 8.3 shows what you might expect to see through the eyepiece.

This chart clearly shows the movement of the asteroid over several hours and also the magnitudes of the surrounding stars but more on that later.

Table 8.1. Selection of asteroids and data from the NASA/JPL Solar System Division 'What's Observable' website

Asteroid		Time			
Number	Name	Rise	Transit	Set	Magnitude
1	Ceres	19:34	00:22	05:10	6.9
6	Hebe	02:31	04:51	07:11	10.8
10	Hygiea	13:08	17:48	22:27	11.7
12	Victoria	17:33	21:08	00:44	11.8
13	Egeria	18:19	00:02	05:45	10.2
27	Euterpe	17:52	22:17	02:42	9.69
30	Urania	19:40	23:21	03:01	10.8
40	Harmonia	15:37	20:30	01:23	10.9
45	Eugenia	22:22	01:25	04:28	11.2
63	Ausonia	17:14	21:51	02:28	11.5
83	Beatrix	00:13	02:23	04:34	11.9
115	Thyra	23:16	00:30	01:45	11
129	Antigone	17:08	21:30	01:51	11.7
192	Nausikaa	17:10	21:53	02:36	11.4
230	Athamantis	18:05	21:21	00:37	11.1
349	Dembowska	20:04	00:18	04:32	10.3
385	Ilmatar	16:53	21:59	03:05	11.5
511	Davida	17:57	23:08	04:18	10.5
654	Zelinda	18:38	20:57	23:16	10.8
925	Alphonsina	20:49	22:45	00:41	11.9

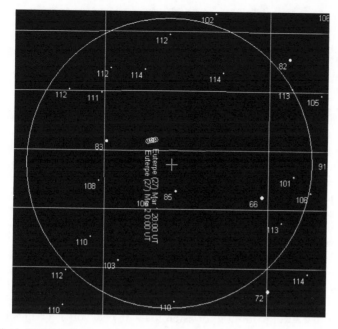

Fig. 8.3. Chart for (27) Euterpe with eyepiece circle, 50 arcsec, overlaid (Credit: *Megastar*, ELB Software).

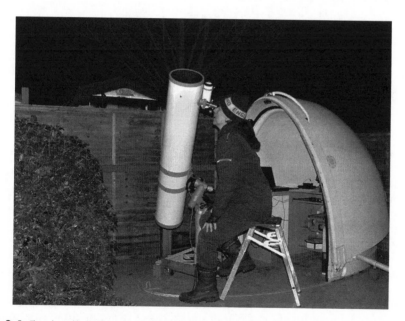

Fig. 8.4. The author and his 10Ð Orion Optics UK Newtonian reflector (Photo by the author).

What to Record

Logbook

Having done all the preparatory work, set up your telescope – Fig. 8.4 shows the author well wrapped up for an evening's viewing – and found your asteroid there are a number of factors you can record. Always keep a record of what you do right from day 1 – it may come in handy when you want to write a book such as this one! Various astronomical organizations produce log sheets or list what should be recorded for each observation, for example, the Astronomical League's Asteroid Club and the British Astronomical Association's Asteroid and Remote Planets Section, and computerized logs are available or you can use your own spreadsheet. At the telescope it is easiest to write your observations in a notebook for later transcribing to your finalized log.

The simplest activity is to record the name and number of the asteroid, instrument used, and the place, date (double date e.g. 20/21 for an observation on the evening of the 20th or morning of the 21st) and time (Universal Time, UT) of your observation, and you can even be rewarded for your efforts. The Asteroid Club section of the Astronomical League, based in the United States, gives awards for observing a specific number of asteroids – 25 gets you a certificate and 100 a certificate and pin.

Drawing the Star Field

To get a little more out of your visual observations you can draw the star field and plot the position of the asteroid. No fancy equipment is needed, just a red torch, pencils and paper with pre-drawn circles, and a clipboard or other firm surface to rest on. When plotting the star field mark the positions of a few of the brighter, widely spread stars first. Then plot the others, trying to keep the correct relationship with all previously plotted stars. It helps to see patterns, e.g., triangles, squares, or straight lines made by the stars. Figure 8.5 is an example of such a plot (translated from hand drawn to computer) of (423) Diotima done by the author on May 24, 2001. It can be seen that the asteroid is just off a straight line extending from stars A and D; stars D, B, and C form a right angled triangle; and stars E and C and the asteroid form an isosceles triangle.

You can make your drawings more interesting by including any deep sky objects such as galaxies or clusters that might be in the same FOV, an example of which is shown in Fig. 8.6. This drawing, of (7) Iris, is a negative version of that made by Eric Graf on May 4, 2008, using a 15 cm (6 in.) Parkes Astrolight Newtonian reflector, magnification ×60, FOV 52′. Graf is indeed fortunate in that he is in reach of a dark site, the Cuyamaca Mountains in southern California, where the naked eye limiting magnitude is around magnitude 6.5, and, with his telescope, he can see down to magnitude 14. Jeremy Perez's 'The Belt of Venus' website carries a wealth of information and guides as to how to draw astronomical objects.

If such an object is particularly colorful, then, for added impact try doing your drawing in color. This is not as hit or miss as it might seem, as the Minor Planet Observer website will give you a list of Asteroid – Deep Space Object (DSO)

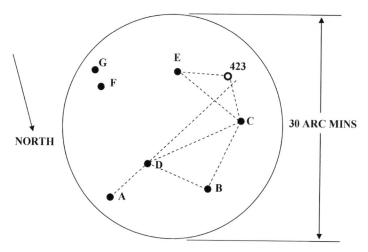

Fig. 8.5. Observation of (423) Diotima (Diagram by the author).

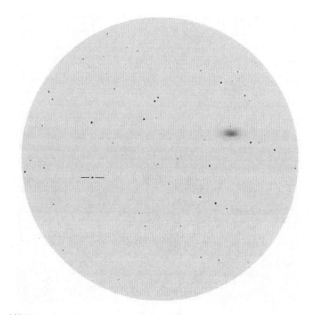

Fig. 8.6. Asteroid (7) Iris near the Sombrero galaxy (M104) (Credit: Eric Graff).

appulses, or close approaches, for any selected month/year. Including other objects will also help to improve your drawing skills – useful if you want to observe deep sky objects in their own right or draw lunar features. Never immediately dismiss something you have seen but can't find on your star chart. It may be that the object has been missed off the chart, but it might just be something new or changed, for example, a new comet or variable star suddenly brightening. However before rushing off and claiming a 'discovery' it is advisable to seek confirmation from a more experienced colleague or your local astronomical organization.

Estimating Magnitude (Visual Photometry)

Having found and plotted the position of the asteroid the next task you might like to try is estimating its brightness or magnitude. This is a useful skill to perfect as it can be used in other fields of astronomy, such as observing meteors, comets, and variable stars. There are various methods for doing this, but all rely on comparing the brightness of the asteroid with that of other stars of known magnitudes in the field of view. Be sure to record which of the methods you have used when reporting observations. Comparing the brightness of the asteroid with all the stars in the FOV leads to a more accurate estimate and highlights any discrepancies between actual and observed magnitudes, which can then be further investigated.

The Fractional Method

This is probably the simplest in practice, and most beginners start in this way. Two comparison stars must be first selected, one just brighter and one just fainter than the asteroid. The brightness of the object under study is then estimated as a fraction of the difference in brightness of the two comparison stars. For example, if the difference between the brighter comparison (star A) and the asteroid is one third the difference between the asteroid and the fainter comparison (star B), then the estimate is written as A(1)Asteroid(2)B. Note that the brighter comparison star is always written first. You might find it easier if the comparisons with the stars that

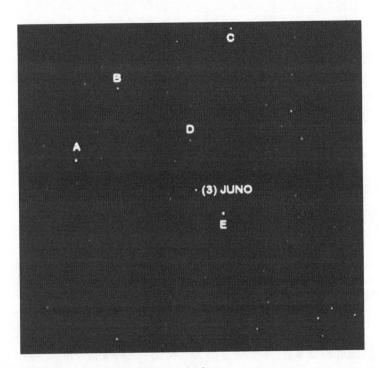

Fig. 8.7. Visual magnitude estimate of (3) Juno (Credit: Maurice Gavin).

are much brighter or fainter than the asteroid are done first. When comparing the asteroid with stars of similar brightness try defocusing slightly, as this will enhance any small differences. In Fig. 8.7 asteroid (3) Juno is indicated between stars D, magnitude 8.7, and E, magnitude 6.2. If you were to estimate that it lies midway in brightness between these two stars you would write E(1)Juno(1)D, e.g., the asteroid is magnitude 7.5 – not a bad estimate, as it was actually 7.8.

The Argelander Step Method

Using a regular step method the observer estimates the difference in brightness between the object and only one comparison star. This process is then repeated several times using other comparison stars. In the Argelander method allowance is made for the unsteadiness of the atmosphere and the imprecision of the human eye. Indeed it could be said that the method relies on these two factors to decide upon brightness steps. An example of the technique follows:

- If, after prolonged viewing, the asteroid appears brighter than the comparison for the same amount of time that the star appears brighter than the asteroid, then the asteroid is assigned the same magnitude as that of the star.
- If the asteroid appears brighter most of the time, but on one occasion it is fainter than the star, then it is noted down as being one step brighter.
- If the asteroid appears brighter most of the time, but on one occasion it is equal to the star, then it is two steps brighter.
- If the asteroid is always brighter than the star but on occasions only just so, then the difference is three steps.

Further steps can also be recorded but are less reliable to estimate. Several comparison stars should be used to achieve maximum accuracy. Note that no attempt is made to estimate the size of an individual observer's step, which actually varies for different observers due to physiological factors. Typical steps fall in the range 0.06–0.09 magnitudes.

Pogson's Step Method

This method is, in fact, relatively popular even though it is probably the most difficult of the three methods described here. Here the observer trains himself or herself to recognize brightness steps of 0.1 magnitude difference. The method demands great discipline from the observer but can be used (with caution) when only one comparison star is present. Where possible, however, several comparison stars should be used.

Estimating Position (Visual Astrometry)

Although positions on the sky can be described in five different ways – equatorial, horizontal, ecliptic, galactic, and heliocentric coordinates – equatorial (right ascension and declination) is the most common and is used throughout this book. One way of estimating the position of an asteroid is to plot a chart using *Megastar* (without the asteroid shown, of course). Using your hand-drawn chart as a guide place the cross hairs where you estimate the asteroid to be, left click the mouse and the coordinates – RA and Dec – will then be displayed at the top of the screen. What you can then do is to repeat the exercise every hour or so and plot the track of the asteroid. The observed movement will verify that you have correctly identified the asteroid.

A word of caution. Visual estimates of positions are not accurate enough for reporting to the Minor Planet Center. Such reporting requires the observer to obtain an Observatory Code and measure positions on images obtained with a CCD camera. This activity is described in Chaps. 10 and 11.

Observing Projects

How Faint Can You Go?

If you live in, or have access to, an extremely dark site, attempting to observe an asteroid without any optical aid would be an interesting project to undertake. Such a project would be simplicity in itself, as all you would need is pencil and paper and perseverance.

Richard Miles, director of the Asteroid and Remote Planets Section of the British Astronomical Association (BAA), undertook such a project from his home in rural Dorset, England, in late 2006 – his target being (7) Iris. Having obtained the asteroid's approximate position (from the *Handbook of the BAA*) he used Norton's star atlas to build up a mental picture of the stars within 3–4° of that position. Moving outside and seated in a comfortable chair he made a mental note of the stars in that area and then returned indoors and sketched what he had seen. If, like most people, you don't have a photographic memory you might find it easier to make a drawing while observing. Viewing through a cardboard tube might help to you to concentrate on the required field of view. For example a 1-in. (2.5-cm) by 12-in. (30-cm) tube will yield a FOV of approximately 5°. Miles found that some (brighter) stars could be seen all the time but others (fainter) only intermittently. Comparing his drawing against the star atlas, *Uranometria*, and the planetarium program, Guide, he concluded that he had managed to observe stars down to magnitude 7.4 and had identified (7) Iris, which he estimated to be magnitude 6.9.

French amateur astronomer Gérard Faure traveled into the mountains to a height of 1,200 m (4,000 ft) to observe the Leonid meteor shower in 2006 and also managed to detect asteroid (7) Iris at magnitude 6.9 without any optical aid.

Following in the Footsteps

By visually observing asteroids you are following in the footsteps of such eminent astronomers as Guiseppe Piazzi, Baron Von Zach, and Heinrich Wilhelm Olbers. You might like to emulate the observations of these early astronomers. There are over 300 asteroids to choose from, as the first to be discovered photographically, by Wolf in 1891, was (323) Brucia. Similar projects have been tried in other areas of astronomy, e.g., recreating the circumstances relating to famous paintings or photographs.

Such an observing project might include:

- Use of a similar instrument. (If any of the original telescopes still exist and you could get access to them, then you really could recreate the discovery circumstances.)
- Observing from or close to the location of the discovery.
- Observing on the same date as the original discovery or when the asteroid is in the same part of the sky (unlikely that the two factors would coincide, but it is worth researching).

Early observations (corrected for precession) are available on the AstDyS website, and an ephemeris for any particular date can also be generated on that website. Be aware that precession will need to be taken into account if you are using actual observational data recorded at the time of discovery.

If you are of a mathematical bent you might like to try calculating orbits using the method developed by Karl Gauss. You could of course just plug the coordinates into an application such as Project Pluto's *Find_Orb* – not so much of a challenge, though!

Conclusion

Timing occultations, Chap. 15, and involvement in the Magnitude Alert Project, Chap. 14, present opportunities for the visual observer to make a useful contribution to our understanding of asteroids. Whatever form of visual observing you under-take do share your results with others via your local and national astronomical groups. By doing so you will encourage others to investigate the path you have taken and meet amateur astronomers who can help you, should you wish, to go down the road described in the following chapters.

Chapter 9

Webcam and DSLR Imaging

Although webcams and Digital Single Lens Reflex (DSLR) cameras are usually pointed at deep sky objects (DSOs), planets, and the Moon they can also be used to image asteroids, and this chapter explains their usage and demonstrates what can be achieved.

Both of the imaging methods described in this chapter can capture asteroids as faint as magnitude 10–11, so suitable targets can be found, as described in the previous chapter in the section 'Targets for Tonight.' More interesting images can be obtained if two or more asteroids are included, several images are stacked to show the movement of the asteroid, or the asteroid is close to a deep sky object such as a galaxy or star cluster. Your local astronomical society will almost certainly welcome the results of your efforts, and the more creative images may find a place in astronomical magazines.

Webcam Imaging

A webcam is a small, cheap, low resolution digital camera originally intended to be used for personal communication over the Internet. It typically operates at about 30 frames/s and is very light sensitive. Initially, amateur astronomers modified such cameras by replacing the standard lens with an adapter so that it could be fitted to a telescope in place of the eyepiece. Webcams, for black and white or color imaging, have subsequently been developed specifically for attaching to astronomical telescopes and are marketed by, for example, Celestron, Meade, Orion, and Imaging Source. These dedicated astronomical packages include improved hardware and software allowing longer exposures of many seconds to be obtained.

Webcam imaging can be compared to visual observing. The best images can be selected from sessions of many hundreds and stacked automatically using software such as *Registax* or *Astrostack* or by doing so manually. Such stacking reduces noise and other artifacts, improving the look of the final image. In a similar manner the visual observer will wait for moments of good seeing to add more detail to their drawing.

Naturally the webcam needs a computer to control it and store the images obtained (DSLRs can also be computer-controlled as mentioned later in this chapter). Those observers who do not have a permanent observatory with the facility to

R. Dymock, *Asteroids and Dwarf Planets and How to Observe Them*,
Astronomers' Observing Guides, DOI 10.1007/978-1-4419-6439-7_9,
© Springer Science+Business Media, LLC 2010

Fig. 9.1. Computer 'dry box' (Credit: Chris Hooker).

protect their computer from the elements can follow the example of UK amateur, Chris Hooker, and construct a wooden 'dry box' with a transparent lid and front access (see Fig. 9.1).

Two examples of the use of webcams by amateur astronomers Chris Hooker and John Sussenbach are described in the following paragraphs.

Asteroid (6) Hebe by Chris Hooker, UK

(6) Hebe is a relatively bright asteroid (tenth magnitude on the nights in question) and, plotting positions from the *Handbook of the British Astronomical Association* in *Uranometria* showed that it would be near a group of ninth magnitude stars (RA: 14 h 46 m, Dec: +06° 50') on the nights of May 23 and 24, 2009, making for an interesting image.

Figure 9.2 (approximately 27 × 18 arcmin, 4 arcsec/pixel) is a composite of 13 separate images (north up and east to the left) taken at 30-min intervals over the two nights and clearly shows the motion of the asteroid relative to the stars. The asteroid is moving from left to right, position angle (PA) 272.9°, at 0.52 arcsec/min. PA is measured from north via east (anti-clockwise in Fig. 9.2). The faintest stars visible are between 12th and 13th magnitude.

Following are some of the technical details.

Locating the Asteroid

- The asteroid was located visually using an eyepiece with cross-hairs or a CCD frame.
- The eyepiece was replaced with the webcam.
- The asteroid was centered in the image and brought into focus.
- The brightness of the image was adjusted using the webcam controls.

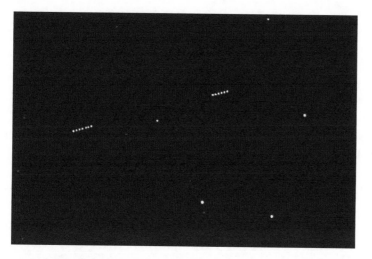

Fig. 9.2. Asteroid (6) Hebe (Credit: Chris Hooker).

Fig. 9.3. Chris Hooker's Williams Optics refractor, Imaging Source camera on a Vixen mount (Credit: Chris Hooker).

Image Capture

- The images were captured in AVI format with a William Optics ZS-66 semi-apo refractor, focal length 388 mm on a Vixen GP-DX mount, using an Imaging Source DBK 21AU04.AS camera with Baader UV/IR rejection filter (see Fig. 9.3).
- Sets of around 40×8.1-s exposures were taken every 30 min (approx), starting at 22:45 UT on each night.
- Dark frame images with the same exposure time were recorded at the start of each session and combined using *Registax* to make a composite dark field image.

Image Processing Part I

- Asteroid images were inspected with the *VirtualDub* video editing program, trailed frames deleted and the remainder saved as a new AVI file.
- The remaining frames were registered, stacked, sharpened and dark frames subtracted, again using *Registax* – dark frame subtraction being necessary to keep the sky background as dark as possible and eliminate hot pixels.

The result was a set of 13 images over two nights, each showing the asteroid in a different position relative to the background stars.

Image Processing Part II

- Image orientation was measured in *Iris* using the point-spread function tool to determine the positions of two reference stars, and the gradient of the line joining them calculated.
- Images from the second night were rotated to the same orientation using the Rotate function in *Iris*.
- Images were combined using the Mosaic function in *Iris*, which gives the option of choosing the brightest pixel value at each point in the overlap region. This ensures the asteroid images appear at the correct brightness (stacking in *Registax* results in the brightness being averaged, so the asteroid brightness is reduced by a factor equal to the number of stacked images).

The Imaging Source camera used in this example has the capability to take longer exposures than 'true' webcams that, as mentioned earlier, typically operate at several or several tens of exposures per second.

(1) Ceres – John Sussenbach

When the dwarf planet (1) Ceres reached opposition on February 15, 2009 – 15 days after perihelion – it was, at 1.5832 AU from Earth, closer than it will be for the next 1,000 years. Dutch amateur astronomer John Sussenbach recorded this close pass during a period of good seeing the previous day.

The technical details were:

Image Capture

- The images were obtained with a DMK 21AF4.AS monochrome planetary camera attached to a Celestron C11 with a Televue ×3 Barlow lens and Astronomik R(ed)G(reen)B(lue) filters (see Fig. 9.4).
- 980 × 1/5 s. exposures were taken through each filter with the gain and gamma set to 100% and 40%, respectively.

Image Processing

- The best 500 frames per color were stacked with *Registax 4.0*.

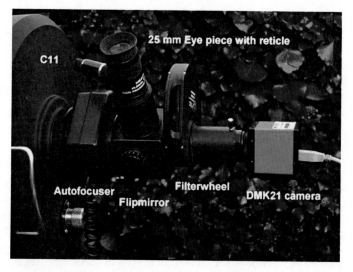

Fig. 9.4. John Sussenbach's Celestron C11 and imaging attachments (Credit: John Sussenbach).

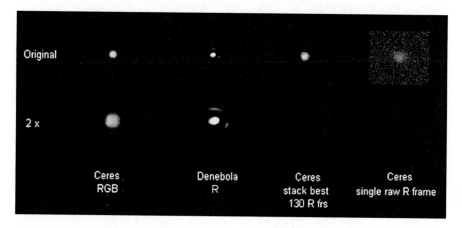

Fig. 9.5. Dwarf planet (1) Ceres (Credit: John Sussenbach).

The final image, Fig. 9.5, clearly shows Ceres as a tiny planet rather than a point-like asteroid. At this time the angular diameter subtended by the dwarf planet was approximately 0.83 arcsec. The limb darkening and spherical shape of (1) Ceres shown in the stacked color image at lower left in Fig. 9.5 is clearly evident when compared with the seeing disk of the star Denebola (β Leonis), imaged at the same time.

This may be the first time that an amateur astronomer has imaged and successfully resolved the disk of (1) Ceres. The Hubble Space Telescope image, Fig. 4.8, may show more detail, but then it is somewhat larger and better placed! Those wishing to try to emulate John may like to make a note of the following close approach dates of (1) Ceres: December 18 and 20, 2012, February 1, 2018, March 21, 2023, and January 9, 2027. Of these dates Ceres will be at its brightest on December 18, 2012, and, at magnitude 6.73 it may even be visible to the naked eye when seen in a very dark sky.

An asteroid will be at its brightest when three factors coincide – it is at perihelion, at opposition (a perihelic opposition), and at zero or a very small phase angle. As described in Chap. 14 an asteroid undergoes a surge in brightness – the opposition effect, typically 0.3–0.5 magnitudes – under such conditions due to multiple reflections occurring on its surface.

DSLR Imaging

DSLR cameras can be mounted on most tracking mounts, piggybacked on a telescope, or can use the telescope itself as a gigantic telephoto lens. Figure 9.6 shows a DSLR camera on a barn door or Scotch mount. The drive motor turns so as to move the upper platform at a sidereal rate (the drive screw can be operated manually for simplicity and economy). A curved drive screw will improve tracking accuracy.

Focusing a DSLR camera at night is not as simple as doing so in the day as, usually, there is nothing bright enough for an autofocus lens to adjust to. The big advantage of this type of camera over a film version is that you can instantly see whether or not your image is in focus. There are a number of ways of achieving proper focus, two of the simpler ones being:

– Using a focusing mask as shown in Chap. 10, Fig. 10.6
– Trial and error by taking an image and then zooming in on it in play-back mode

Being uncooled, DSLR camera images are inherently noisier than CCD versions, but image noise (simply random, unwanted signal) can be reduced by stacking a number of short exposures. Taking, and subsequently subtracting, dark frames will also improve the quality of your images. Dark frames are obtained with the lens covered but with all other settings the same as those used for obtaining the actual images. Chapter 10, the 'CCD Basics' section, has more to say on this subject.

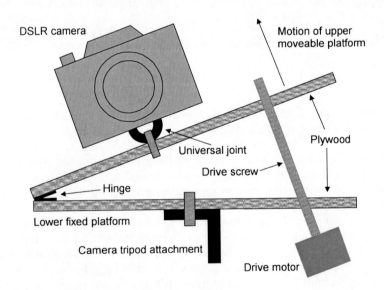

Fig. 9.6. DSLR camera on a barn door or Scotch mount (Diagram by the author).

As can webcams, DSLR cameras can be computer-controlled, and a number of commercial (e.g., *Astroart*) packages are available.

Most DSLR cameras store images in JPEG format, but, for much improved quality in terms of color and detail, Raw format is the best option. The downside of this format is that images cannot be viewed on the camera but, when imported into a computer, allow greater control over how they are processed to produce a better final result. Some cameras allow you to take images in both formats but apply the camera settings in terms of image manipulation only to the JPEG version.

What can be achieved is next illustrated by the work of two amateur astronomers, Michael Clarke and Maurice Gavin (the latter better known for his spectroscopy work).

(44) Nysa – Michael Clarke

Figure 9.7 shows two images of asteroid (44) Nysa taken 24 h apart. During that time the asteroid has moved approximately 8 arcmin. The faintest stars in these images are approximately magnitude 13, while the asteroid itself is magnitude 10.

The images, 30-s exposure time, were obtained using a Canon 350D DSLR camera attached to an 80-mm f/7 refractor on a Vixen GP equatorial mount – see Fig. 9.8. A 30-s exposure on Michael's 120-mm f/8 doublet refractor will show stars down to magnitude 15.

Asteroids (11) Parthenope and (16) Psyche: Maurice Gavin

A more unusual configuration is that constructed by UK amateur astronomer Maurice Gavin. Figure 9.9 shows his camera on its gravity drive mount together with an explanatory diagram. By balancing the camera and counterweight very

Fig. 9.7. Images taken on February 21 and 22, 2007, showing movement of asteroid (44) Nysa (Credit: Michael Clarke).

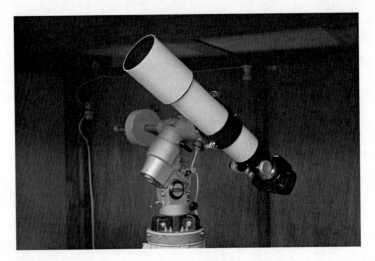

Fig. 9.8. Michael Clarke's 80 mm refractor, Canon camera, and Vixen mount (Credit: Michael Clarke).

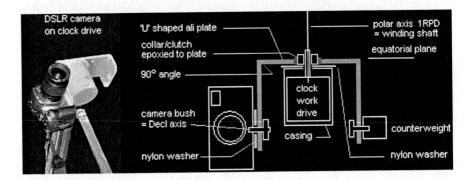

Fig. 9.9. Canon 300D DSLR camera and gravity drive (Credit: Maurice Gavin).

slightly in favor of the camera (with the camera 'west' of the meridian and the counterweight to the east) gravity alone was sufficient to operate the clock, giving a controlled 'fall' to track with the stars. Operated in this mode the drive requires no power and is thus free from mains/battery power or leads to trip over. It was stored under wraps on his patio and could be brought into action in a matter of moments. The only maintenance required was a bi-annual spray of WD40.

With care the mount accurately tracked a DSLR with a 135 mm fl lens for about 2 min. Typically a Jupiter 85-mm fl f/2 Zenit lens reached magnitude 11 in 30 s – sufficient for imaging the brighter asteroids. A collection of fixed-focal length Pentax screw/M42 thread lenses could be attached to a Canon 300D (Rebel) camera using a Kood M42 to EOS adapter.

Due to Maurice's proximity to London, sky pollution can be onerous. On images taken with a digital color camera the sky shows as a hideous yellow–brown. Although special filters can reduce the problem they have downsides and are not totally effective. His solution was simple. The selected raw image was copied (using any popular image processing software) and heavily blurred (Gaussian Blur) to 'remove' all the stars. The resulting image was subtracted from the original,

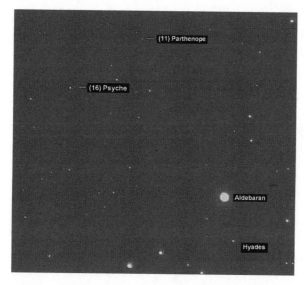

Fig. 9.10. (11) Parthenope and (16) Psyche. Image obtained with an 85-mm focal length f/2 lens attached to a Canon 300D (Rebel) DSLR camera (Credit: Maurice Gavin).

turning the yellow–brown sky to a pleasant neutral gray. This image could then be further enhanced by subjecting it to modest contrast stretching and sharpening. Figure 9.10, an image of asteroids (11) Parthenope and (16) Psyche in the constellation of Taurus, was obtained at 20:42 UT on January 4, 2006. The asteroids were magnitude 10.5 and 10.0, respectively, and stars down to approximately magnitude 12 appear in the image.

Conclusion

This short chapter demonstrates what can be achieved with webcams and DSLR cameras, the work of John Sussenbach being particularly notable. However, for making astrometric and photometric observations to the required standards it is a common practice to use a CCD camera as described in the following chapters.

Chapter 10

Astrometry Tools and Techniques

In this chapter we will:

- Define astrometry and its uses.
- Describe what you need to perform accurate astrometric measurements (to better than 1 arcsec).
- Say how you can obtain an Observatory Code from the Minor Planet Center (MPC), which qualifies you to submit astrometry to that organization.
- Explain the art of tracking and stacking (to find faint and/or fast moving asteroids).

The 'Guide to Minor Body Astrometry' on the MPC website is a MUST READ. In addition Appendix D of this book – 'Astrometry How-To,' contributed by Tim Spahr, director of the MPC, may help you avoid many of the common astrometric pitfalls. More advanced methods will be mentioned in the following chapter. This chapter is very much a 'getting started' exercise.

The What, Why, and How

What Is Astrometry?

Astrometry in its broadest sense is the measurement of positions, parallaxes, and proper motions of an astronomical body in the sky. In the next two chapters we will use a narrower definition – the measurement of the position of an asteroid in terms of its Right Ascension (RA) and Declination (Dec), known as equatorial coordinates. For nearby objects such as asteroids and in particular near-Earth objects, these coordinates can vary slightly depending on your location on Earth. Hence the importance, when obtaining an ephemeris from the MPC, for example, of specifying one's exact position, which includes latitude, longitude, and height. If you have an Observatory Code, as described later in this chapter, then inputting that will supply these parameters.

R. Dymock, *Asteroids and Dwarf Planets and How to Observe Them*,
Astronomers' Observing Guides, DOI 10.1007/978-1-4419-6439-7_10,
© Springer Science+Business Media, LLC 2010

Why Is It Needed?

Accurate astrometry is necessary to:

- Determine the orbit of a newly discovered asteroid.
- Refine that orbit, leading to that asteroid being numbered.
- Help prevent loss and assist recovery of lost asteroids.
- Support radar observations (which can determine the size and shape of an asteroid).
- Improve the accuracy of occultation predictions.
- Determine asteroid masses.

How you can partake in these activities will be covered in Chap. 11.

How Is It Done?

In the days of astrophotography measurement of position was a complicated and time-consuming task involving the use of a mechanical plate-measuring engine; a single measurement could take several hours to complete. Nowadays it requires a CCD image, an accurate star catalog, an astrometry software package, and a PC; measurement time is reduced to minutes if not seconds. The accuracy of the position so determined will, to a large extent, only be as good as the accuracy of the reference stars used. *Astrometrica* allows you to select from several catalogs, which it accesses via the Internet – USNO-B1.0, NOMAD, and UCAC 2 are recommended for astrometry.

A brief mention of video cameras before moving on. Accurate astrometry of fast-moving objects, usually near-Earth asteroids, is difficult with a CCD camera due to timing problems. Although not an ideal tool for astrometry a video camera may well give more accurate results in such circumstances. A guide to video observing techniques can be found on the Minor Planet Center website.

Tools of the Trade

Here we describe the author's set-up. Other set-ups are covered in Chap. 7. Together they should give you some good ideas as to how you might like to proceed. Mine started out as a visual observatory, so it may not be untypical for newcomers to the world of the amateur astronomer. When I began observing I had a vague notion that I would proceed from visual observations to photographic imaging. After several failures with film and with affordable CCD cameras becoming more readily available I abandoned the former for the latter.

My very first telescope was a cheap disaster, my second was a very good, very rugged TAL 10 cm (4 in.) reflector, and my third and present instrument is a 25 cm (10 in.) Newtonian reflector. The latter coupled with a Starlight Xpress MX516 CCD camera allows me to reach approximately magnitude 15 with a 30-s exposure. By stacking the images, which will be described in the next chapter, I can reach fainter than magnitude 18. So I would suggest that a 25 cm (10 in.) reflector is a good starting point, but, if you can afford it, then the larger the better. Naturally the larger the telescope the more expensive it will be, and you will need a larger, more costly observatory to house it.

The advantage of a permanent, polar aligned set-up is that it is ready to use at a moment's notice. This is particularly relevant to imaging newly discovered asteroids (those appearing on the MPC's NEO Confirmation Page, for example) and even more so for very fast moving objects (VFMOs), as will be described in the next chapter.

– A schematic of the author's set-up is shown in Fig. 10.1. It includes:
– A reflecting telescope on a go-to equatorial mount controlled by *Megastar* via a Skysensor hand controller.
– A CCD camera controlled by *Astroart*. You will need a CCD camera with an FOV of at least 10×10 arcmin or you may find there are too few reference stars on the image. Mine is 10×8 arcmin, and I do have to check the number of stars in the FOV when choosing targets, as sometimes there are too few.
– A laptop PC.
– A GPS receiver and *TAC32* software (to input time correct to at least 1 s and determine latitude, longitude, and height of site above sea level).
– A CD R/W drive to save images.

Here are a few telescope do's and don'ts – **do** ensure your telescope is collimated and **don't** clean the mirrors unless absolutely necessary.

Note on filters – for astrometry alone no filters are required, but for photometry at least a Johnson V filter is necessary for your measurements of magnitude, to conform to a well known standard.

Your set-up may well be very different but, in summary, you will need: a telescope on a motorized mount, a means of controlling it, a CCD camera, imaging software, an accurate time source, and a large amount of storage for your images. You will notice an aerial in Fig. 10.1. This connects an observatory laptop via a wireless router to a PC in my study, allowing me to control matters from a warm room once I have everything set up.

Astrometry Tools and Techniques

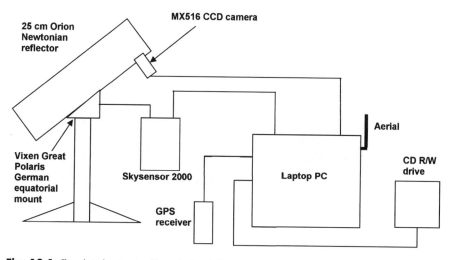

Fig. 10.1. The author's observing set-up (Diagram by the author).

CCD Basics

A CCD camera consists of a matrix of picture elements, called pixels, that convert the incident light into electrons, which are then fed to a computer and converted into an image. It is perhaps worthwhile just dwelling for a few moments on how pixels convert photons into electrons (see Fig. 10.2), as this will help to understand the meaning of, and the need for dark frames, flat fields, and image calibration.

Each pixel generates electrons that are fed to your computer in the form of electric current to form an image. The number of electrons generated by each pixel depends on the amount of incident light (photons), and the always present dark current. The output of each pixel is measured in Analog Digital Units (ADUs) – the maximum, full-well or saturated value usually being 65,536. To obtain a true representation of the area of sky imaged, the dark current must be subtracted and the variation in number of electrons generated in each pixel by the same quantity of incident photons allowed for. Calibrating the images – applying dark frames and flat fields – will perform the necessary corrections. Although this might sound complicated it is quite easy to do, or rather imaging processing software will do the job for you. Note that the CCD camera should not be moved until all images and calibration frames have been obtained and that the latter should be obtained during each imaging session.

A dark frame is obtained by taking an exposure of the same duration as the image while preventing any light falling on the CCD. This can be done quite simply by fitting a light-tight cap on the end of the telescope tube (see Fig. 10.3). Best practice is to take a number of dark frames, at least five, and combine them. Photometry software packages such as *AIP4WIN* and *Canopus* have facilities to do this.

A flat field is slightly more complicated in that you need to take an image of a uniformly illuminated surface. There are a number of ways of doing this, but a 'light bin' seems to give satisfactory results ('bin' because it was made from a plastic

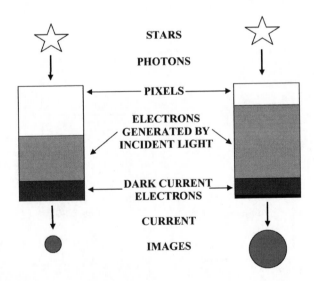

Fig. 10.2. Pixels – from light to image (Diagram by the author).

Fig. 10.3. Light-tight cap for obtaining dark frames (Photo by the author).

Fig. 10.4. Light bin for obtaining flat fields (Photo by the author).

waste bin, as shown in Fig. 10.4). More detailed diagrams of the light-tight cap and light bin are shown in Chap. 12. Images are obtained with the telescope pointed at the illuminated surface – a minimum of 3 s exposure is recommended, and the ADU value of any pixel should not exceed around 50% of maximum. Imaging software such as *Astroart* will provide the required image statistics telling you whether you have under or over exposed the flat field image.

Pixel Size

Perceived wisdom is that a resolution of approximately 2 arcsec/pixel gives satisfactory results in terms of astrometric (and photometric) accuracy. What you have, or can afford, in the way of a telescope or CCD camera may govern the result. For example when I started CCD imaging I already possessed a 10 in. (25 cm) Newtonian reflector, and my pockets were only deep enough for the cheapest Starlight Xpress CCD camera. You may be able to compensate for too high a resolution by binning – whereby the output from several pixels is combined electronically.

You can do the mathematics to calculate the CCD field of view (FOV) and resolution (using my equipment as an example):

1. First calculate the plate scale, which is (206,265)/Focal length of the telescope in mm = 206,265/1,626 = 126.9 arcsec/mm.
2. Then calculate the size of the CCD chip from pixel size (0.0098×0.0126 mm) and number of pixels on the chip (500×290) = 4.9×3.6 mm.
3. The FOV is therefore ((126.9×4.9)/60) × ((126.9×3.6)/60) = 10.4×7.6 arcmin.
4. The resolution is the plate scale divided by the number of pixels/mm = 126.9/89 = 1.4 arcsec/pixel.

Obtaining an Observatory Code

This section explains how to obtain an Observatory Code, including selecting the target asteroids, the basics of imaging, and the measurement of those images. The same procedure can also be applied to the projects described in the following chapter.

Why Do You Need One?

Observatory Codes are issued by the MPC when they are satisfied with the accuracy of your astrometry. Having obtained such a code you can then submit astrometric data to them.

Choosing Asteroids to Image

How to obtain your Observatory Code is illustrated by the procedure I followed towards the end of the year 2000 to obtain that code for my home observatory. As suggested on the MPC website I selected six asteroids with numbers in the range of 400–3,000. Using *Megastar* I chose asteroids that were in the southeast to ensure that they would be visible for a number of nights and thus avoid having to restart the exercise should I experience a period of cloudy skies (perusing your long-range weather forecast before starting this exercise may help you to avoid such a trap). It also helps to reduce time spent moving between asteroids if they are as close together as possible.

Asteroids, between magnitudes 13 and 16, chosen were:

(3259) Brownlee	Discovered by J Platt at Palomar in 1984
(403) Cyane	Discovered by A Charlois at Nice in 1895
(1040) Klumpkea	Discovered by B Jekhovsky at Heidelberg in 1925
(1719) Jens	Discovered by K Reinmuth at Heidelberg in 1950
(1468) Zomba	Discovered by C Jackson at Johannesburg and L Boyer at Algiers in 1938
(591) Irmgard	Discovered by A Kopff at Heidelberg in 1906

Choosing brighter asteroids (in CCD terms) allows the exposure times to be kept fairly short and therefore minimizes tracking errors.

Imaging the Asteroids

Here is a method I have used, frowned upon by some. I move my telescope to the required coordinates using *Megastar's* go-to facility with a CCD framing eyepiece installed. I then replace (very carefully so as not to move the telescope) the eyepiece with the CCD camera. The camera and drawtube are marked so that the camera is installed in the correct orientation. A stop ring or mark on the draw tube aids focusing, which may need a slight adjustment after the first images are obtained. I focus by eye by taking short exposure time images and adjusting the focus manually, as I would if an eyepiece were installed, until the images of stars are as sharp as possible. This should suffice for the whole imaging session unless, for example, the temperature changes significantly. Figure 10.5 show the Starlight Xpress MX516 camera mounted on my telescope. The stop ring can be seen above the focusing knob.

To assist focusing a mask (see Fig. 10.6) can be placed over the end of the telescope similar to the placement of the light-tight cap. The mask is made from cardboard with three holes, making an equilateral triangle, cut out. If the image is out of focus each star will be depicted by a cluster of three dots. The focus is adjusted until these three dots merge into one.

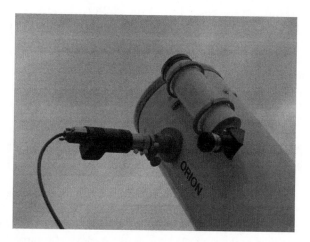

Fig. 10.5. Starlight Xpress CCD camera with stop ring in position (Photo by the author).

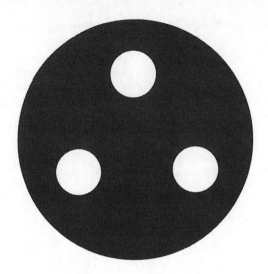

Fig. 10.6. Focusing mask (Diagram by the author).

The six asteroids were each observed over a period of 1 h on two separate nights. It took approximately 3 weeks to complete all the observations. Having aligned the telescope on the target asteroid a typical observation run proceeded as follows. (Note that over the years I have amended the process I used at first, to reflect how I would do it now):

- The PC was set to the correct time using a GPS receiver connected to the laptop via a USB port. (If you have an Internet connection handy then a clock sync program can be downloaded.)
- The asteroid name was entered and numbered into the FITS header. (Images are always saved in FITS format.)
- A number of test images were taken to verify tracking, focus, and that the pixels were not saturated. A maximum of two thirds of the full-well value for any pixel is recommended. The maximum exposure time is also limited by the ability of your telescope to track accurately and the motion of the asteroid. As a rough rule of thumb the maximum exposure time to avoid the image of the asteroid trailing is 4/motion of asteroid in arcsec per min (see more on this under 'Tracking and Stacking' below).
- The *Astroart* software was set to Continuous, 5 min delay between exposures, and Autosave image as a FITS file. This allowed all the image data to be saved including the date and time of the exposure. The MPC requires that the mid-points of the exposure times are submitted, so check that your imaging software is assigning this time rather than the start time of the exposure, for example. Your astrometry software may allow you to input whether the time in the FITS header is start, middle, or end of exposure and, from this, calculate the mid-time. The delay between images ensures the PC clock resets to the correct time if the software you use freezes the clock while the image is downloaded to your PC.
- I imaged for a period of 1 h. (After the first few images were obtained I checked to ensure that they had been saved correctly – well worth doing.)
- Dark frames were obtained. At this time I did not obtain flat fields, as I was not planning to include magnitudes in my report.

A go-to feature on your telescope certainly helps, as you can image one asteroid and fairly easily switch to the next, image it, and so on. (Remember that each asteroid must be observed on two separate nights.) If you have a Newtonian reflector try and stay away from the meridian so that you don't have to reverse the telescope as it passes that point.

Processing the Images

- From the dozen or so images obtained for an asteroid during each of the two evenings it was observed, I selected three images taken roughly 15 min apart.
- Each image was processed by subtracting the dark frame (if you are also measuring magnitude then apply a flat field, but don't process the images in any other way).
- The position of the asteroid was measured using *Astrometrica* and the USNO-B1.0 catalog. Tutorials are available on the *Astrometrica* website, and it is quite easy to use. Briefly:

  - Set up the configuration file.
  - Load dark frame (and flat field for photometry).
  - Load images.
  - Carry out astrometric data reduction (if you are unsure as to where the asteroid is in your image compare it with a star chart – *Astrometrica* circles the stars so picking out the asteroid should not be too difficult). Alternatively you could use the blink facility. Figure 10.7 shows the Object Verification window which appears when you click on the asteroid in your image. Entering the asteroid designation and clicking on Accept will add the relevant data to the MPC Report File. Figure 10.8 shows an image after astrometric data reduction – the asteroid is now circled and numbered.

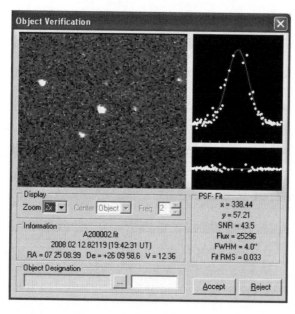

Fig. 10.7. *Astrometrica* Object Verification screenshot (Credit: *Astrometrica*).

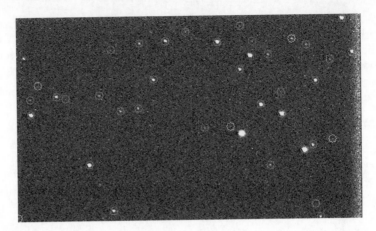

Fig. 10.8. Image after astrometric data reduction (Credit: *Astrometrica*).

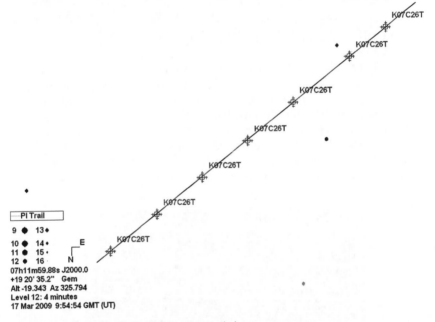

Fig. 10.9. Astrometry plotted using *Guide* (Credit: Guide-Project Pluto).

Verification of Results

It is wise to check one's results prior to submission to the MPC. This can be done by comparing your astrometry with an ephemeris from the Minor Center for the time of observation. Another way, which will show up any inconsistencies, is to plot your results in a planetarium program such as *Guide*. Printing out the plot and simply drawing a line through the positions (see Fig. 10.9), will show if there are any outliers that should be ignored or remeasured.

Submission of Results to the Minor Planet Center

There is a very specific format that MUST be used to report observations to the MPC – an example report is shown below. Software such as the widely used and recommended *Astrometrica* will produce reports in the required format. Whatever software you use do ensure your observations are reported in plain ASCII format (plain text, not, for example, HTML format). Each line in the body of the report is 80 characters long, so set your e-mailer to automatically break lines at a slightly larger number.

COD XXX
OBS R.Dymock
MEA R.Dymock
TEL 0.3-m f/5.9 reflector + CCD
ACK MPCReport file updated 2008.12.09 14:45:21
AC2 roger.dymock@ntlworld.com
NET USNO-B1.0

00941	C2008 12 07.70371 07 28 46.59 +17 25 54.9	15.8V	XXX
00941	C2008 12 08.66792 07 48 38.62 +17 43 33.8	15.4V	XXX
00941	C2008 12 08.67245 07 48 44.10 +17 43 36.8	15.4V	XXX
00941	C2008 12 08.67695 07 48 49.64 +17 43 41.5	15.2V	XXX
00941	C2008 12 08.68144 07 48 55.27 +17 43 45.3	15.2V	XXX

The lines above the actual observations start with a code describing the information included in that particular line:

COD – observatory (would be XXX when submitting your astrometry to obtain this code, as would be the code at the end of each line of observations)
OBS – the observer
MEA – the measurer
TEL – description of the telescope used
ACK – will enable the MPC to automatically acknowledge receipt of your observations
AC2 – e-mail address to which the MPC will respond
NET – catalog used

Each line of observational data includes asteroid number, C for CCD observation, date, RA, Dec, magnitude, band (V in this case), and XXX, as you do not yet have an observatory code. Magnitude should only be included if you are confident that this is accurate to at least ±0.1 magnitude. The measurement of this will be covered in Chaps. 12 and 13.

The first time observations are reported additional information, which may be submitted using the COM prefix on each line, is required by the Minor Planet Center, for example:

– Postal address
– Observatory name and site
– Observatory position: longitude, latitude, height above sea level, and source of coordinates. These can be obtained using a GPS receiver. For more accurate

coordinates take the average values over a long period of time. (The MPC now recommends using Google Earth.)

– Details of telescope set-up

A few days after submitting my measurements to the MPC I received a short email to the effect 'Your site is now observatory code 940.' I also received a slap on the wrist for submitting the observations in the wrong format – but that only goes to show that the MPC staff are quite helpful, as they didn't reject my submission but translated it to the correct, ASCII or plain text, format.

An Observatory Code is specific to an observatory, so if you use another telescope at a different site you will need to ensure that it has another Observatory Code. When using robotic telescopes check if this is so and how observations should be submitted to the MPC, e.g., who is the observer and who is the measurer. As an example the Skylive and Sierra Stars organizations allow observations to be submitted using their observatories with the user as the named observer and measurer.

Tracking and Stacking

If you were imaging a faint star you could merely take a long exposure or add together (stack) several shorter ones without having to take into consideration whether or not the target was moving. Unfortunately asteroids do move, and some of them move quite quickly, so this factor must be taken into account. The matrix of pixels accumulating photons in a CCD camera can be compared to a similar matrix of buckets catching rainwater in a field. The more rain that falls, the more water there is in the buckets and, likewise, the longer your exposure the brighter will be the object on the image.

Now imagine a very small rain cloud moving so quickly across the field that it deposits only a few drops of water in each bucket. No matter how long you leave the buckets in position (or how long an exposure you take) you are not going to collect any more water in each bucket (or image fainter asteroids). However, knowing the speed and direction of movement of the cloud you could run after it with your bucket and collect more water (and image fainter objects), and this is exactly what the Track and Stack facility in *Astrometrica* allows you to do.

Figure 10.10 shows one of a sequence of images of asteroid 2004 BZ$_{57}$, obtained on August 25, 2005. Stars are visible but not the asteroid.

Astrometrica allows you to load a number of images, weed out any unsatisfactory ones due to poor tracking, for example, and then stack them allowing for the motion of the asteroid. Having loaded a sequence of images, 57 × 45-s exposures in this example, and selected the Track and Stack option, you are then able to enter the asteroid motion in terms of speed and position angle as shown in the *Astrometrica* screen shot, Fig. 10.11. In this instance the necessary data was obtained from the MPC Minor Planet Ephemeris Service web page.

Having aligned the image with the catalog overlay, USNO-B1.0, the asteroid shows as a single point and the stars as lines of images (see Fig. 10.12). The position of the asteroid can then be determined as previously described. The motion data can be varied and the images restacked if the asteroid shows signs of trailing.

Astrometrica has been and will be mentioned many times in this book. In the view of many amateur and professional astronomers there is no better astrometric

Fig. 10.10. Single image of asteroid 2004 BZ$_{57}$ (Image by the author).

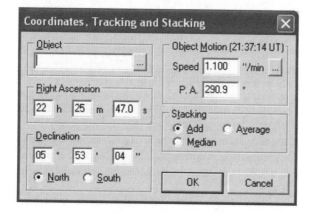

Fig. 10.11. *Astrometrica* Track and Stack input window (Credit: *Astrometrica*).

Fig. 10.12. Stacked image of Asteroid 2004 BZ$_{57}$ (Credit: *Astrometrica*).

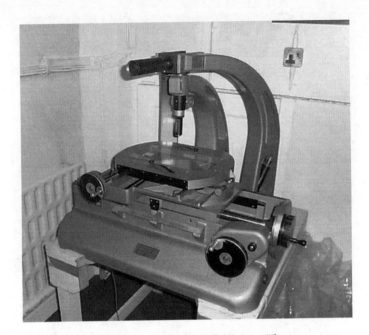

Fig. 10.13. Plate-measuring machine (Credit: Peter Birtwhistle, Great Shefford Observatory, UK).

software. A single positional measurement takes seconds, but it wasn't always so. Back in the late 70s and early 80s a mechanical X-Y plate-measuring machine (see Fig. 10.13) had to be used, and what can now be done in seconds would then have taken hours.

Conclusion

Do keep a record of your observing sessions – any problems you experienced, how you overcame them, and how you processed your images.

Astrometry really isn't too difficult. All you need are clear nights to obtain the images and the time and patience to process them.

So now that you know the basics of astrometry the next chapter will describe the many uses to which that skill can be put.

Chapter 11

Astrometry Projects

Details of asteroids for which astrometry is required can be obtained from a number of sources as described in this chapter, which is structured so that the easier options, imaging the brighter asteroids, are described first followed by the more complex. A plea – these sources will provide you with enough targets to keep you busy so do not waste your time and that of the Minor Planet Center by imaging and reporting on those asteroids for which data is not requested. As with most things, walk before you run, and progressively 'push the envelope' to see what you can achieve. Naturally the equipment required becomes more complex and more expensive as you climb the tree of difficulty. All observations should be sent to the Minor Planet Center in the standard format.

Examples of projects being carried out by more advanced amateur astronomers complete the chapter.

Follow-Up Observations

Why are they necessary? There are numerous reasons for observing newly discovered asteroids as well as those that have been known for some while. For example, many asteroids discovered by surveys done by professional astronomers, but also the reasonable number found by amateurs, can easily be lost if their orbits are not accurately computed. To do this requires observations over many nights and, if the asteroid is to be numbered (see Chap.2), over several orbits or oppositions. In addition, if an asteroid has not been observed for a number of years, its actual and predicted positions might differ by several arc minutes. These and other observational requirements, described in this chapter, present us with numerous targets. Amateur astronomers generally have more time, and easier access to telescopes, than professionals to make the necessary observations and can thus perform a very valuable role here.

For target selection, here is a summary of what you need to know:

- Decide which area of your sky is most suitable for observing.
- Check the approximate RA and Declination of that area for the period of observation.
- Choose a target using the websites described later in this chapter.

R. Dymock, *Asteroids and Dwarf Planets and How to Observe Them*, Astronomers' Observing Guides, DOI 10.1007/978-1-4419-6439-7_11, © Springer Science+Business Media, LLC 2010

- Obtain the latest orbital elements and/or an ephemeris from the Minor Planet Center's Minor Planet and Comet Ephemeris Service.
- Plot the track of the target using a planetarium package such as *Megastar* or *Guide*.

Lowell Observatory

This observatory's Hierarchical Observation Protocol (HOP) will generate a list of asteroids based on the capabilities of the user's set-up and various other requirements, such as angular distance of the asteroid from the Sun and the Moon. You can then select targets based on asteroid group and reason why observations are required. In this example we selected Main Belt objects (MBOs) and mass determination, and the results are shown in Table 11.1.

This latter selection criterion identifies asteroids that will have a close encounter with other, more massive objects, such that observations will aid in determining the masses of the smaller objects. The close pass of a large body by a smaller one will slightly change the orbit of the latter, and accurate astrometry can define the new orbit. These asteroids are well within the capabilities of quite modest equipment and thus a good starting point.

In addition to the HOP, Lowell Observatory also maintains a Critical List of Asteroids whose orbits can be improved by further astrometric observation.

Spaceguard

The Spaceguard Foundation was set up in Rome on March 26, 1996. One of its purposes is "to promote and coordinate activities for the discovery, pursuit (follow-up), and orbital calculation of the NEO at an international level." The asteroids it deems worthy of observation can be found on the Spaceguard Priority List. This list classifies the need to observe near-Earth asteroids into four categories: urgent, necessary, useful, and low priority and focuses on newly discovered objects to ensure that the highest possible percentage of these bodies can be recovered at

Table 11.1. Asteroids selected for observation using HOP

Asteroid name	R.A.	Dec.	V Mag
(00014) Irene	14 04 19.27	+01 23 35.9	9.0
(00413) Edburga	11 09 22.15	+28 13 16.9	16.1
(00090) Antiope	10 19 33.83	+13 10 01.8	14.3
(01216) Askania	12 42 33.96	+10 19 40.8	16.1
(06325) 91EA1	10 23 48.07	+20 04 36.5	16.2
(00538) Friederike	11 32 32.95	+08 46 31.5	15.2
(00351) Yrsa	14 54 51.81	−03 44 56.9	12.7
(00201) Penelope	10 43 12.30	+10 28 00.8	13.7
(02873) Binzel	12 02 50.26	+10 23 26.1	16.1
(00112) Iphigenia	10 38 58.70	+06 29 10.7	14.6
(02195) Tengstrom	10 25 55.90	+16 00 56.8	16.9
(03002) Delasalle	11 16 18.72	+14 35 47.2	16.1
(00076) Freia	11 49 22.57	+00 21 18.7	13.3
(00908) Buda	10 36 07.20	+27 47 20.8	14.8

Table 11.2. Spaceguard Priority List

Priority	Object	Inserted in this category	R.A.	Dec.	Elong.	Mag.	Sky uncert. in arcsec	End of visibility
Data for 2009 Apr 28, 22:00 UT								
LP	2009 DL1	2009 Mar 29	15 h 06 m	−34.0	158	16.8	0	2009 Sep 13
US	2009 HV44	2009 Apr 26	16 h 03 m	+00.8	151	17.1	1	2009 Dec 24
LP	2008 WN2	2009 Apr 6	14 h 04 m	−17.7	174	17.2	0	2009 Jul 7
LP	2009 DE47	2009 Apr 2	13 h 53 m	−24.7	167	17.4	0	2009 Sep 29
UR	2009 HG60	2009 Apr 28	12 h 54 m	+05.9	150	17.6	12	2009 May 23
NE	2009 EC	2009 Apr 24	15 h 31 m	−76.5	117	17.9	4	2009 Nov 20
US	2009 HD	2009 Apr 21	14 h 03 m	−23.7	169	17.9	2	2009 Jun 16
LP	2009 CQ1	2009 Apr 12	13 h 14 m	+28.2	134	18.2	0	2009 Jul 13
LP	2009 CR2	2009 Apr 2	09 h 02 m	+45.2	86	18.6	0	2009 Nov

Table 11.3. Orbital elements of 2009 HG_{60} as at June 18, 2009

Epoch	2009 June 18.0
Mean anomaly, M	35.22580
Semimajor axis, a	1.9961274 AU
Eccentricity, e	0.6473767
Argument of perihelion	283.46046°
Longitude of the ascending node	212.98394°
Inclination	6.97132°
Absolute magnitude, H	22.7
Slope, G	0.15

future apparitions. *Astrometrica's* 'Track and Stack' feature can be considered the 'medium' for making faint objects, invisible on a single image, suddenly appear!

The Spaceguard Priority List can be sorted by Priority, Object, Magnitude, and Right Ascension. Table 11.2 shows a portion of the list sorted by magnitude. Asteroids on this list are typically fainter than magnitude 17, with the odd brighter one, but those with modest equipment will usually be able to find something suitable to image.

From this list you might select 2009 HG_{60}, depending on whether its position and magnitude suit your observing site and equipment capabilities. Querying the Minor Planet Center's Ephemeris Service will provide the necessary data, Table 11.3, to plot the asteroid track using *Megastar*.

Follow-Up Astrometric Program (FUAP)

Much attention is paid, and rightly so, to near-Earth asteroids due to the danger they pose to our very existence. However if the less glamorous Main Belt asteroids are not observed for several years they may be lost due to small but continuous changes in their orbits. The FUAP run by the Italian Organization of Minor Planet Observers, a section of the Unione Astrofili Italiani (UAI), was set up to identify such objects. Asteroids are listed by the number of oppositions at which they have been observed and, within those categories, a class indicating the urgency with which observations are required. Most of these asteroids are magnitude 17 or 18 and thus require a medium to large telescope (10 in. or larger reflector).

Table 11.4. Selection of asteroids from the FUAP (partial data)

Minor planet	Class	R.A.	Dec	Mag
2007 VV187	A	15.44	+05	18.5
2007 XP17	A	14.86	+00	17.8
2007 YW3	A	15.82	−52	17.1
2007 YO6	A	15.20	−14	18.3

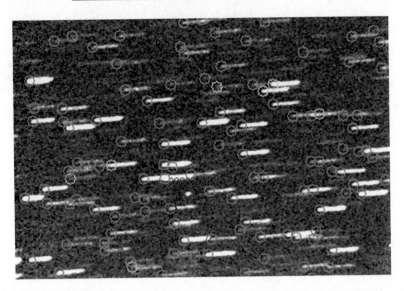

Fig. 11.1. Stack of 15 × 20 s exposures showing asteroid 2006 NM. The *circles* indicate the positions of the reference stars (Credit: *Astrometrica*).

Obtaining the data required to point your telescope is a simple matter of selecting those asteroids in terms of magnitude and coordinates deemed suitable for your set-up and location, and then selecting 'Get ephemerides/orbits.' For example on May 1, 2009, asteroid 2007 YW$_3$ was listed as a 'minor planet observed at only one opposition and not observed for 1 year,' and in the 'A', Very Urgent, class (see Table 11.4). Its southern declination, −44°, makes it most suitable for southern hemisphere observers or northern hemisphere observers with access to robotic telescopes located south of the equator.

On October 13, 2005, this author imaged FUAP asteroid 2006 NM. Figure 11.1 shows a stack of 15 × 20 s exposures, and the asteroid can be seen as a single point just below center. The 13th lived up to its name as the telescope was being operated remotely, and a slight adjustment of the position of the telescope to try and center the asteroid on the image caused a telescope reversal, though no damaged equipment.

Minor Planet Center

There are several pages on the MPC website that you can use to produce lists of asteroids to image:

- The MPC's NEA Observation Planning Aid
- Dates of Last Observation of NEOs
- Dates of Last Observation of Unusual Minor Planets

Objects on these lists are typically magnitude 19 or fainter, although a small number of magnitude 17/18 objects are listed, and thus require the use of a large telescope, 14 in. (35 cm) to 16 in. (41 cm) aperture or larger. You can customize these pages to suit your circumstances by choosing, for example, the area of sky within which you wish to observe, magnitude of objects, object motion, object type, and your Observatory Code.

Discovery Confirmation

When an asteroid is first discovered it is given a designation by the discoverer, e.g., 8V98296. The discovery is reported to the Minor Planet Center using the format described in the previous chapter and listed on the Near-Earth Object Confirmation Page (NEOCP). Upon confirmation of that discovery, by astrometry from a further night's observations, for example, it is assigned a provisional designation by the MPC, e.g., 2006 NM. Objects appearing on this page are, by virtue of their motion or orbit, likely to be NEOs but may also be comets.

The NEOCP can be accessed and suitable objects chosen to suit one's location and set-up. These objects are usually faint – typically magnitude 18–20, and therefore a large telescope, such as a 16 in. (41 cm) Schmidt–Cassegrain, is necessary to image them. If you have a smaller telescope, 10 in. (25 cm), for example, your observing opportunities will probably be limited to a few brighter objects per month. Orbits calculated from a few observations are somewhat uncertain, so the sooner an object appearing on the NEOCP, Table 11.5 is an example, can be followed up the better the chance of imaging it, and the wider the field of view of one's CCD camera the better.

Object 9J3EF4B was of particular interest, as it was bright, as objects on this page go, and its ephemeris showed that it was moving quite slowly for an NEO – 10–11 arcsec/min. The ephemeris for this object was obtained from the MPC and the predicted path displayed in *Guide* (see Fig. 11.2). The overlaid CCD frame is approximately $30' \times 20'$ and gives an indication of the amount of time the object will take to cross the CCD field of view.

An alternative method is to obtain the observations for the NEO of interest, load them into Project Pluto's *Find_Orb*, and generate the orbital elements and an ephemeris. The orbital elements from either source can be loaded into a planetarium program such as *Megastar* and the predicted track displayed.

In addition an uncertainty map showing the nominal location (0 RA and Dec offsets) and a range of other possibilities based on different assumptions used to calculate the objects' orbits can be accessed (see Fig. 11.3).

Table 11.5. Example NEOCP listing

ID	Added	R.A.	Dec.	V mag
9J9386D	2009 May 13.2 UT	14 26.0	−07 37	19.9
9J93460	2009 May 13.2 UT	13 11.6	−06 30	19.9
SW40gk	2009 May 13.1 UT	15 59.0	+11 51	19.7
9F3EF4C	2009 May 13.1 UT	12 11.2	+55 55	17.1
9J92AEB	2009 May 13.1 UT	07 45.4	+20 28	20.8
9J3EF4C	2009 May 12.3 UT	12 12.1	+55 53	16.9
9J3EF4B	2009 May 12.3 UT	14 36.3	+47 41	16.5

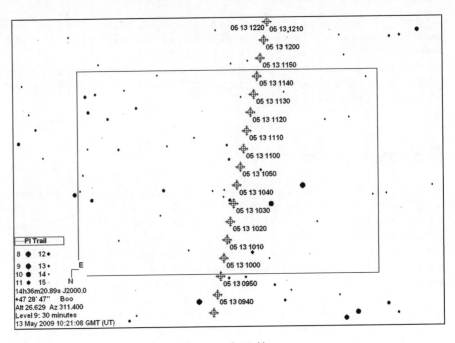

Fig. 11.2. Predicted path of NEOCP object 9JEF4B (Credit: Project Pluto-Guide).

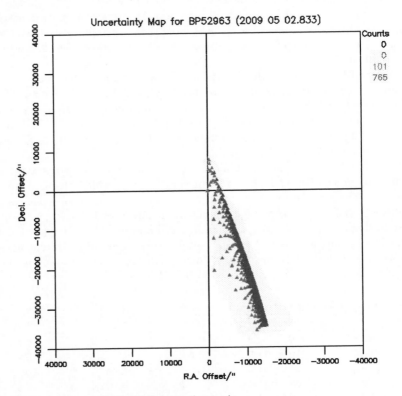

Fig. 11.3. Uncertainty map for NEO BP52963 (Credit: Minor Planet Center).

The spread of these predictions will help you decide whether or not to attempt to image the object and where to search in relation to the nominal position. UK amateur Peter Birtwhistle's search and imaging of NEOCP object AU52949 was described in Chap. 6.

To avoid the asteroid trailing on the image, and thus affecting the accuracy of your astrometry, the exposure time needs to be short, especially for fast-moving objects, requiring the use of *Astrometrica's* 'Track and Stack' facility to detect the object and carry out astrometry. A rough formula for calculating exposure times was mentioned in the previous chapter, but a more accurate formula is:

Exposure time in seconds = 60 × (image scale in arcsec/pixel)/(motion in arcsec/min)

Using this formula will ensure that your exposures are short enough to avoid trailing but as long as possible to improve the signal to noise ratio. 9J3EF4B was moving at a little under 11 arcsec/min and therefore the exposure time, to avoid trailing and assuming an image scale of 2 arcsec/pixel, would be 12 s. When choosing an object from the NEOCP it is worthwhile examining the motion to ensure that the object will stay in the field of view (FOV) long enough to enable a reasonable number of images to be obtained before having to reposition your telescope. In the case of 9JEF4B it would have stayed in the same FOV for just under 2 h. Even with a relatively small FOV, 12 × 8 arcsec, you would be able to image it for around 30 min without moving your telescope. So the opportunities, although relatively few, are there for observers with smaller, e.g., 8–10 in. (20–25 cm) telescopes.

Sometimes a very faint asteroid can be difficult to locate on a stack of images, so here is one way of detecting such an object. Assuming you have taken, say, 40 short duration images, stack all of them and also stack them in sets of 10. Blink all 5 stacks. The object may then be easier to see in the stack of 40, and you might then be able to locate it in each of the stacks of 10. Do make sure the object in the total stack is in a position appropriate to the other stacks.

Observations should be reported to the MPC in the usual way, using the same ID for the object as given on the NEOCP but with some urgency – e.g., the same night as you make your observations. In the previous chapter mention was made of checking one's observations by plotting them using *Guide*. Another method is to compare your observations with those made previously using Project Pluto's *Find_Orb*. By loading both sets of observations into that application and computing the orbital elements you can calculate the residuals for your data. Residuals? If an orbit is calculated from all observations of that object the residuals are the differences between the measured and calculated positions. You can check the residuals relating to your own astrometry on, for example, the NEODyS website (for specific observations) and the MPC website (annual summaries for each Observatory Code). You should aim to be consistently less than 1.0 arcsec.

If an object that appeared on the NEOCP is subsequently confirmed to be a near-Earth asteroid a Minor Planet Electronic Circular (MPEC) will be published that will include discovery and follow-up observations, observer details, and orbital elements. These can be viewed on the MPC website or, by subscription, received by e-mail.

Asteroid Discovery

Don't believe what you may read about the age of amateur discovery being over, that the professional sky surveys make all the discoveries and the like. Not so! Amateurs continue to discover new asteroids – 4,021 in 2008 for example. If you think you have made a discovery, then there are a few rules to follow:

- Read and comply with the relevant sections in the MPC's 'Guide to Minor Body Astrometry,' the main points of which are included here.
- Make at least two observations of the object at least 30 min apart.
- If at all possible image the object on the following night, as this will increase the likelihood of a formal designation being assigned.
- Check against known objects using the MPC's MPChecker, NEOChecker, and NEOCMTChecker facilities.
- Use the MPC's NEO Rating page to determine the probability that your observations are indeed of an NEO.
- Report the object to the MPC in the approved format, assigning your own ID.

If the object is subsequently deemed to be an NEO it will be listed on the NEOCP. The time taken for this to happen may be minutes or hours, so be patient. The MPC will send you a list matching your ID to a provisional or permanent designation. Assuming you have made a discovery, wait for between 7 and 10 days before reporting further follow-up observations.

A word of caution. Do be very careful when claiming a discovery, making every effort not to tread on anyone's toes. It can be a bit of a minefield, and people can be very sensitive about these matters. The 'path' from discovery to naming is described in Chap. 2.

It is always worthwhile accessing the MPChecker while processing your images, just to see what else you might have caught in addition to the particular object you were imaging. In January 2009 this author imaged NEO 2009 AD$_{16}$, which I had selected from the Spaceguard Priority List mentioned earlier in this chapter. MPChecker showed that the Main Belt asteroid (MBA) 102509 should be in the same field, and indeed I had captured it. In Fig. 11.4 the NEO is the almost vertical line of dots (produced by stacking several images), and the MBA is indicated just to the right of that line. Although they appear to be close together they were actually 8.5 million miles apart.

An 8 or 10 in. (25 cm) reflector, or larger, plus a CCD camera, plus a great deal of patience, should enable you to make discoveries of Main Belt asteroids. Assuming that the typical MBA moves at 0.5 arcsec/min then an exposure time of 4 min would be short enough to prevent trailing of the asteroid image. Avoid using a filter, as this will improve the signal-to-noise ratio and enhance your chances of detecting fainter objects.

As most of these are close to the ecliptic, approximately within plus or minus 20°, one approach would be to image a series of adjacent, or slightly overlapping, fields in that region and then repeat the process twice with a 30-min delay between each image. Since many asteroids brighten significantly close to opposition it may improve your chances if you concentrate on an area at solar longitude 180°. To be sure of getting something useful for your efforts select an object using the various resources mentioned previously and check the images for additional objects. You can, by searching well away from the ecliptic, avoid the areas covered by the major professional surveys by accessing the Sky Coverage Plot on the MPC

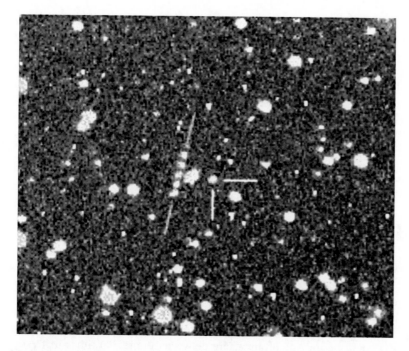

Fig. 11.4. NEO 2009 AD$_{16}$ 'passing' MBA 102509 (Credit: *Astrometrica*).

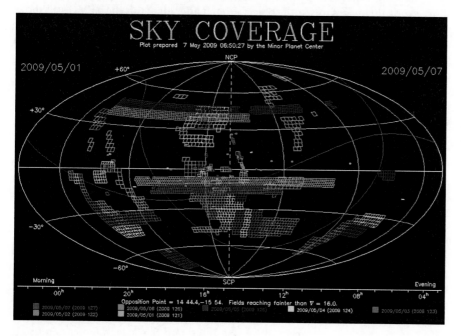

Fig. 11.5. Sky coverage by the major surveys (Credit: Minor Planet Center).

website – Fig. 11.5 is an example for the week May 1–7, 2009. It might seem odd to suggest attempting to discover asteroids around the time of the full Moon, but this is a period the major surveys avoid, so perhaps searching then, but as far away from the full Moon in the sky as you can, is another ploy worth considering.

After calibrating your images (applying flat fields and dark frames) the images can be blinked using *Astrometrica's* Moving Object Detection facility, for example. Any object that moves in an even manner in more or less a straight line across your images will be detected but should be confirmed by examining the images manually. If the various MPC checking routines show no known objects in that area, then your object may indeed be a discovery. As mentioned previously try to image the object on the following night. Finding it again should not be too difficult, as an 0.5 arcsec/min object will only have moved 12 arcmin in 24 h.

One final point – do check your images carefully, as that newly discovered 'asteroid' might have a tail and therefore be a comet, as was the case with 2002 EX_{12} described in Chap. 2.

Advanced Amateur Astrometry

Very Fast Moving Objects (VFMOs)

As their name suggests VFMOs are fast moving, usually very small, objects that are only bright enough to be discovered by the survey telescopes when close to Earth. Whereas the motion of a typical NEO might be of the order of 5 arcsec/min or less, a VFMO could be moving at anywhere between 10 and 100 times that rate. Your exposure times will need to be very short, and you will need to stack many of these short-exposure images to 'see' the object. Even if after all of this the VFMO still shows as a short trail on the image it is still possible to obtain astrometry by measuring each end of the trail. The time for each measurement must be adjusted, as the time for the beginning of the trail will be mid-exposure time minus half the exposure duration and for the end of the trail will be mid-exposure time plus half the exposure duration (make sure you know which way the object was moving!). Also bear in mind that astrometry packages may automatically center the curser over the brightest part of the asteroid image, which will probably not be the end of the trail.

NEO 2007 RS_1, Fig. 11.6, was moving at 244 arcsec/min and can be seen as a 16 arcsec trail on this stack of 40×4 s exposures taken on September 4, 2007.

Accurate timekeeping by a GPS receiver or via the Internet, using freeware such as *Dimension 4* for example, is an absolute essential. The amateur astronomer attempting to image such asteroids must therefore pay close attention to the NEOCP to determine, by their fast acceleration, which are indeed VFMOs. You should try to image them as soon as they are announced, as the uncertainties in their positions grow rapidly. (It has been said that VFMOs find you rather than the other way around!) If you are fortunate enough to 'trap' one, then try to position your telescope ahead of the object to obtain further images. This can be done by eyeball extrapolation of the orbit, by using orbit and ephemeris calculation software such as Project Pluto's *Find-Orb*, or by accessing the object's ephemeris on the NEOCP.

Interlaced stacking, a variation on the Track and Stack technique, is a useful method developed by Peter Birtwhistle to help identify faint VFMOs, particularly in crowded star fields. In this method two stacks of images are used with each stack comprising alternate images, e.g., images 1, 3, 5, 7, etc., in one stack and images 2, 4, 6, 8, etc., in the other (see Fig. 11.7). When these two stacks are blinked the two positions of the moving object are close together and thus easier to spot.

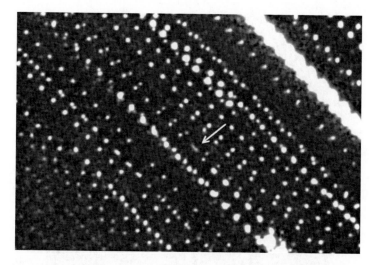

Fig. 11.6. NEO 2007 RS$_1$, seen as a short trail (Credit: Peter Birtwhistle, Great Shefford Observatory, UK).

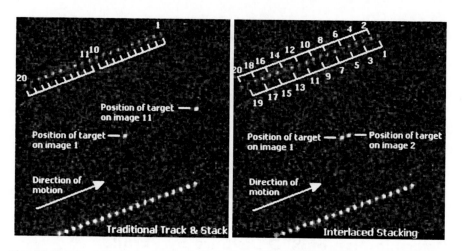

Fig. 11.7. Traditional vs. Interlaced stacking (Credit: Peter Birtwhistle, Great Shefford Observatory, UK).

Not all objects on the NEOCP turn out to be natural objects. Orbiting observatories and spacecraft passing close by have also made an appearance here (see Fig. 11.8).

Fortunately help is available to determine if there are likely to be artificial satellites in your images – *IDSat* and *Sat_ID* software packages both serve this purpose, as does the MPC's Distant Artificial Satellites Observation Page. Actually satellites such as Geotail and IMP8, ephemerides obtainable from the aforementioned MPC page, make good VFMO practice targets.

It is also worth noting that the magnitude of VFMOs may change quite significantly due to their rapid passage past Earth, rotation, and change in phase angle – moving from 'full' to 'new' or vice versa.

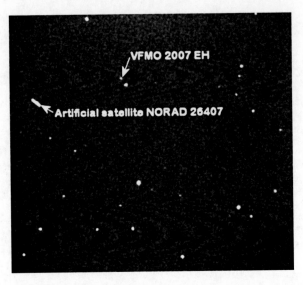

Fig. 11.8. A VFMO and an artificial satellite (Credit: Peter Birtwhistle, Great Shefford Observatory, UK).

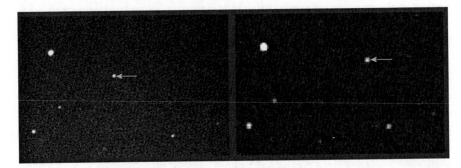

Fig. 11.9. Images of NEO 2005 CR$_{37}$ showing parallax effect (Credit: Peter Birtwhistle, Great Shefford Observatory, UK/Monty Robson).

Determination of NEO Distance by Parallax

It is possible to determine the distance to an NEO with a high degree of accuracy using triangulation. Such an exercise was carried out between 2004 and 2006 by Peter Birtwhistle (UK) and Monty Robson (US), based on techniques developed by US student Lisa Doreen Glukhovsky, for which she won an Intel Foundation Young Scientist Award in 2003. Figure 11.9 shows images of NEO 2005 CR$_{37}$ taken at the same time by Monty Robson (left) and Peter Birtwhistle (right). The displacement of the NEO in comparison to the background stars can be clearly seen. These images highlight the necessity of correctly reporting your own location in observations sent to the MPC, or your astrometry will be in error.

From the astrometry obtained simultaneously by the two observers and knowing the distance between their locations, it was possible to calculate the distance to the asteroid (see Fig. 11.10). The mathematics is a little too complicated to explain here, involving as it does geographic, cylindrical, Cartesian, and equatorial coordinates. Two points worthy of note – the longer the baseline the better, and observatories located in the same time zone offer more opportunities for simultaneous imaging.

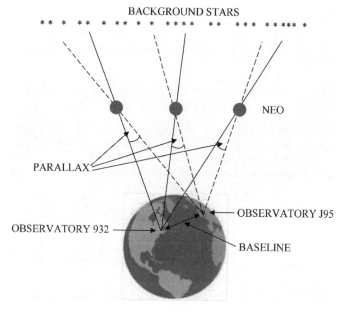

BACKGROUND STARS

NEO

PARALLAX

OBSERVATORY J95

OBSERVATORY 932

BASELINE

Fig. 11.10. Calculation of distance by simultaneous imaging of NEO (Diagram by the author).

Moving Object and Transient Event Search System (MOTESS)

MOTESS is operated by US amateur astronomer Roy Tucker at his Goodricke–Pigott Observatory in Tucson, Arizona, and is primarily used to search for near-Earth objects. This is quite a unique observatory in the world of amateur astronomy, so it is worthy of some description. Roy's first discovery was the Aten asteroid 1997 MW_1 and his most recent Amor asteroid 2008 SP_7; he also searches for comets and makes variable star observations. The telescopes are three home-built 14 in. (35 cm) f/5 Newtonians on a short yoke English mounting (see Fig. 11.11), housed in a conventional roll-off roof observatory. These are fixed in position and scan the same strip of sky each night, thus no expensive, precision-drive mechanism, computerized go-to, or auto guiding is required. The cost of the instrument is in those components that are required for the imaging process – primarily the optics and the cameras.

The home-built CCD cameras, 1,024 × 1,024 chips with 24 μm pixels produce, in combination with the telescope system, an image scale of 2.83 arcsec/pixel and a FOV of 48.3 × 48.3 arcmin. In front of each camera is a filter slide mechanism that permits insertion of clear, V, or I filters. Consideration was initially given to the use of neutral density filters to attenuate the light levels during times of a bright Moon, but it was realized that color filters produce the desired attenuation and also provide useful color photometry. The filters are arranged so that image triplets are ordered in time – I, V, and I – permitting interpolation of an I magnitude measurement at the time that the V magnitude measurements are made.

The fully automated system is operated in drift-scan or time delay integration (TDI) mode such that the image drifts across the CCD camera mounted at the focal plane of the telescope at the same rate as the camera outputs its data to the attached PC.

Fig. 11.11. Moving Object and Transient Search System (MOTESS) telescopes (Credit: Roy Tucker).

This mode of operation will be explained in a little more depth in Chap. 15, which covers occultations of stars by asteroids. The use of undriven telescopes limits observations to regions close to the celestial equator. At more polar declinations star images no longer drift across the CCD in straight lines but begin to move along arcs, producing star images elongated in the north–south direction, making accurate astrometry impossible.

In normal operation, the three telescopes are aimed at the same declination but spread in Right Ascension at intervals that can be varied between 15 and 60 min to produce a data stream of image triplets separated in time, which can be analyzed to reveal moving objects (asteroids and comets) and time-varying objects (variable stars). The same region of sky, a strip 48 arcmin wide at a particular declination, limited at the west and east ends by evening and morning twilight, is searched continuously every clear night without any observer participation or movement of the telescopes. The advantage over using a single telescope is that the latter has to be continuously moved in order to scan the same strip of sky three times and the area of sky that can be covered in a night is much greater with a three-telescope system. Because of their relatively slow motion along a primarily east-west line Main Belt asteroids are generally observable on multiple nights. Identification of asteroids usually requires three images of a star field to unambiguously detect the moving object and avoid false detections due to, for example, 'hot' pixels and cosmic rays.

This set-up will produce 1–1.5 gigabytes of image data per clear night (150–220 image triplets, each covering 0.64 square degrees). Such a large amount of data cannot possibly be examined by eye, and so it is necessary to rely upon computer image processing to search for interesting objects. *PinPoint* software, which can search an entire night's images in 2–3 h, is used to automatically find moving objects and generate astrometry reports for the Minor Planet Center.

During a period of almost 3 years, this simple, inexpensive system has reliably collected high-quality science images with minimal manual intervention and downtime other than periodic optical cleaning and re-evacuation of camera cryostats. Tucker invites others to consider the construction of similar instruments as a possible solution to their need for a low-cost source of astronomical data for research and classroom instruction. There is value in participating in a collaboration operating a large number of these instruments, so inquiries regarding such cooperative ventures would be very welcome.

An Outer Solar System High Ecliptic Latitude Survey

Irish amateur astronomer Eamonn Ansbro is conducting this survey at his Kingsland Observatory with the objective of determining the inclination distribution of objects in the Edgeworth–Kuiper Belt.

As described in Chap. 3, Classical EKBOs fall into two classes – those described as dynamically 'cold' at low inclinations and others, dynamically 'hot' at higher inclinations. Most trans-Neptunian objects have been found by surveys centered on the ecliptic; therefore, the current known population may not be truly representative of the actual total. This project may therefore yield a more representative picture of these distant objects. For this survey a 0.9 m (36 in.) telescope and an Apogee AP8 SITe CCD camera with a FOV of 23×23 arcsec is used. Theoretically it is capable of imaging a Mars-sized object out to 300 AU and a Jupiter-sized object out to 1,200 AU. All the functions of the observatory, telescope pointing, and CCD imaging are automatically controlled using *ACP Observatory Control Software*.

By spring 2009, 127 square degrees up to latitude 45° north had been imaged and several EKBOs and 150 asteroids recovered. Figure 11.12 shows magnitude 21.6

Fig. 11.12. Magnitude 21.6 trans-Neptunian object 2000 CN$_{105}$ (Credit: Eamonn Ansbro).

TNO 2000 CN$_{105}$. Each field is imaged three times for 240 s during the course of one night, and each nightly survey covers 2 square degrees. This survey achieved an average limiting magnitude of 21.7, which bettered the average of 20.5 achieved by the 1.2 m Near-Earth Asteroid Tracking (NEAT) telescope on Mt. Palomar!

The survey also includes a search for a hypothetical ninth planet beyond the EKB based on the search areas identified by Patrick S. Lykawka (see Chap. 3), John Murray, and D. Parkinson. Fields centered on the predicted positions of 44 candidate targets are being observed to greater depth (magnitude 22–22.5) and imaged three times on two nights separated by approximately 6 months.

Conclusion

Assuming your astrometry is accurate and you have submitted your measurements to the MPC in the correct format they may be published, depending on the type of object, in the MPC's Daily Orbit Update, MPECs if they relate to a discovery, and on the NEODys and AstDys websites where they are listed by object and observatory. NEO observations are usually published more or less immediately, but Main Belt and more distant objects may take some time to appear on these websites.

There is probably as much, if not more, computer work than actual observing in all of this. However, since there seem to be far more cloudy nights than clear ones, in the UK at least, you will have something to occupy your time with when you can't get outside and observe!

Astrometry and photometry need not necessarily be considered separate projects, as measurements of position and magnitude can be carried out on the same set of images, although not necessarily at the same time. How to do this and the basics of photometry will be covered in the next two chapters.

Chapter 12

Lightcurve Photometry Tools and Techniques

Lightcurve photometry is an aspect of astronomy where there is ample opportunity for amateurs to make a very significant contribution to our understanding of asteroids – lightcurves having been obtained for less than 1% of numbered asteroids. The more complete the sampling of asteroid lightcurves, the better astronomers can develop theories concerning the origin and dynamics of minor planets.

This chapter describes:

- Why the magnitude of an asteroid varies as it rotates.
- A lightcurve and what it can tell us.
- The two types of photometry and why a new method developed by Richard Miles and the author has significantly simplified this task.
- The equipment you need.
- How to measure magnitude accurately (to better than 0.05 magnitude).
- How to generate a lightcurve.

By taking magnitude measurements of an asteroid at and close to opposition it is possible to produce both a lightcurve and a phase curve from which its absolute magnitude can be determined, but more on that in Chap. 14. This chapter will concentrate on lightcurves.

Asteroid Rotation

How we see the rotation of an asteroid depends on its position and the orientation of its spin axis with respect to Earth and the Sun (see Fig. 12.1).

Assuming that the spin axis of the asteroid does not change its orientation with time, when the asteroid and Earth are at positions A and C we see little or no variation in magnitude. This is because the same face of the asteroid is always pointing towards Earth, as shown in the top row of diagrams in Fig. 12.2. However in positions B and D we would see a significant variation in brightness as the asteroid rotates to alternately show us its long axis and short axis, as indicated in the bottom row of diagrams. The rotational period of an asteroid is typically several hours but in the extreme can be minutes or months. It should be noted that if we observe the asteroid at position A, the Sun, Earth, and asteroid orienta-

R. Dymock, *Asteroids and Dwarf Planets and How to Observe Them*,
Astronomers' Observing Guides, DOI 10.1007/978-1-4419-6439-7_12,
© Springer Science+Business Media, LLC 2010

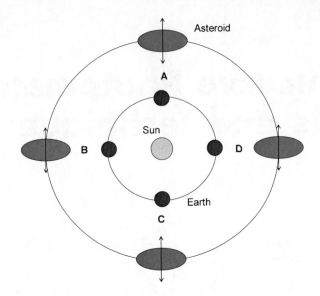

Fig. 12.1. Various orientations of an asteroid's spin axis with respect to Earth and the Sun (Diagram by the author).

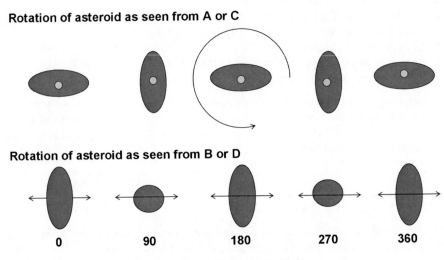

Fig. 12.2. Rotation of an asteroid as seen from various viewpoints (Diagram by the author).

tion shown by position B will not necessarily occur 3 months later. The time between oppositions will depend on the semi-major axis and eccentricity of the asteroid's orbit.

Figure 12.3 is a real-life example showing how lightcurves vary depending on the orientation of the asteroid's spin axis and the Sun–asteroid–Earth relationship. Although the orientation of the spin axis of (755) Quintilla is unknown, Fig. 12.3a

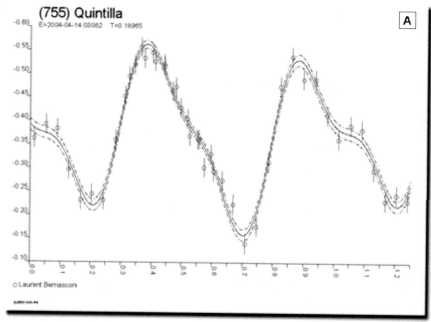

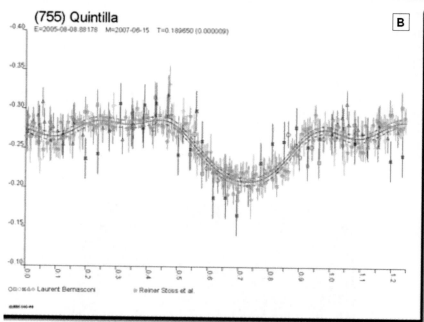

Fig. 12.3. Lightcurves of (755) Quintilla (Credit: Višnjan School of Astronomy).

would represent a situation similar to the bottom row of diagrams in Figs. 12.2 and 12.3b, the top row. The lightcurves were generated by Raoul Behrend from data collected during 2005 by a team of students at the Višnjan School of Astronomy:

Laurent Bernasconi, Petra Korlevic, Maja Hren, Aleksandar Cikota, and Ljuban Jerosimic, led by Reiner Stoss.

An Asteroid Lightcurve

Figure 12.3a shows a typical double peaked lightcurve – the variation in magnitude being due to the 'potato' shape of the asteroid shown in Fig. 12.4 (the shape more formally known as a tri-axial ellipsoid – further described in Chap. 13), and its orientation with respect to Earth and the Sun as shown by the bottom row of diagrams in Fig. 12.2.

The variation in brightness can sometimes be very obvious from the CCD images. Figure 12.5 shows both CCD images and a lightcurve of Apollo class near-Earth asteroid 2001 FE_{90} observed by Peter Birtwhistle June 27, 2009.

What Can a Lightcurve Tell Us?

In addition to the period of rotation the lightcurve of an asteroid can be used to determine:

– Its shape.
– Its composition (fast rotating asteroids, with periods of less than 2.25 h, are almost certainly to be solid bodies – a 'rubble pile' asteroid would fly apart if it rotated so quickly).
– Orientation of the spin axis (but observations over several apparitions are needed to determine this).
– Its size (in conjunction with observations in the infrared).

Fig. 12.4. (243) Ida – a typical 'potato'-shaped asteroid (Credit: NASA/NSSDC Photo Gallery).

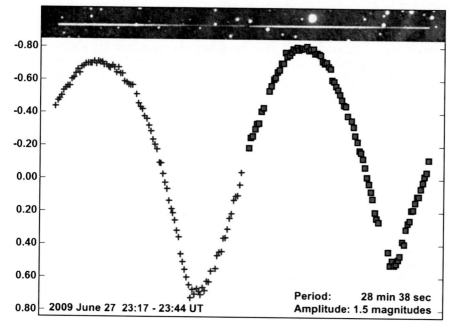

Fig. 12.5. Images and lightcurve of near-Earth asteroid 2001 FE₉₀ (Credit: Peter Birtwhistle, Great Shefford Observatory, UK).

The figure contains the labels:
2009 June 27 23:17 - 23:44 UT
Period: 28 min 38 sec
Amplitude: 1.5 magnitudes

What Is Photometry?

Photometry is the measurement of the brightness of a celestial object. In simple terms the magnitude of an asteroid is calculated by comparing its brightness with that of a comparison star or stars of known magnitude. In the early days when photoelectric photometers were used it was common practice to use a single comparison star, designated C, and a check star, designated K, to ensure that the comparison star itself was not a variable. The object under investigation was designated V for variable (originating from the use of photometry to study variable stars).

Differential Photometry

Differential photometry, as its name implies, is the measurement of the difference in magnitude between the variable and one or more comparison stars on the same CCD image. Changing atmospheric conditions and variations in dimming due to the altitude of the objects under study are likely to affect all in an equal manner and can thus be ignored. Software packages such as *AIP4WIN* make such measurements a relatively easy task. However, deriving actual magnitudes is problematic in that comparison stars with known accurate magnitudes are few and far between, and it is unlikely that you will find many, if any, such stars on a typical CCD image. So to obtain an accurate measure of the magnitude of the target asteroid, one must ordinarily resort to 'absolute' or 'all-sky' photometry.

Absolute or All-Sky Photometry

This approach is however much more complicated than differential photometry in that the sky must be adequately clear (sometimes referred to as 'photometric'), standard stars need to be imaged (usually some distance from the target asteroid and therefore not on the same CCD image), at least one filter must be used, and extinction values – nightly zeropoints, airmass corrections and the like – need to be taken into account in order to transform these measurements on to a standard system such as Johnson–Cousins. Such a transformation is also necessary if several observers wish to compare and integrate their results.

Filters transmit specific wavelengths while blocking others. The original UBV (Ultra-violet, Blue, Visual) photometric standard system was devised by H. L. Johnson and W. W. Morgan in the 1950s. The system was further refined by Johnson himself, Cousins, and Landolt to include two additional bands, R (red) and I (infra-red), and is referred to as the Johnson–Cousins (J–C) UBVRI standard – the most commonly used astronomical photometry system. These wave bands are defined in Table 12.1.

The R(ed), G(reen) and B(lue) filters used for colour astro-imaging are not equivalent to the J-C filters and should not be used as equivalents.

A New Approach

A significant development in this field came about with the publication of a paper by Richard Miles and this author in the June 2009 issue of the *Journal of the British Astronomical Association*. This paper is included in full, as Appendix C, towards the end of this book, but the prior analysis of LONEOS, CMC14, and 2MASS data by John Greaves to derive a formula for calculating V magnitude deserves a mention here. The paper describes how, from a single image or, if necessary to improve the signal-to-noise ratio (SNR), a stack of several images taken through a Johnson V filter, it is possible to obtain the V magnitude of an asteroid accurate to about ±0.05 magnitude or better. All that is required is access to the Carlsberg Meridian Catalog 14 (CMC14) from which data the magnitudes of the comparison stars can be calculated. This method allows for the deriving of V magnitudes of comparison stars and the asteroid under study, in the range 10 < V magnitude < 15. Previously the most accurate source of data was the *Tycho* catalog, but this was only accurate down to approximately magnitude 10.5.

Initially magnitudes were calculated by accessing the CMC14 catalog and performing the necessary calculations using a spreadsheet, but this proved to be a laborious exercise, and some automation was obviously required. *Guide 8.0* was

Table 12.1. Johnson–Cousins wave bands

J–C wave band	Center frequency (Å)	Band width (Å)
U	350.0	70.0
B	438.0	98.5
V	546.5	870.0
R	647.0	151.5
I	786.5	109.0

updated to allow CMC14 star data to be automatically downloaded via the web, enabling suitable comparison stars for any particular asteroid to be identified, but a spreadsheet was still required to calculate the magnitude of the asteroid. The major advance, which makes determination of asteroid magnitudes a simple task compared with absolute or all-sky-photometry, came about after a considerable effort on the part of Richard Miles and Herbert Raab, with the incorporation of the methodology into the latter's *Astrometrica* software. From a single, calibrated image it is now possible to obtain both the position and magnitude of an asteroid to a high degree of accuracy. It is necessary to use different settings within *Astrometrica* for astrometry and photometry, so each image must be processed twice – a small price to pay.

In this chapter we will work through examples of basic differential photometry and photometry using the new approach mentioned above, which is covered in more detail in Appendix C. For those of you wishing to try your hand at traditional absolute or all-sky photometry refer to the excellent book *A Practical Guide to Lightcurve Photometry and Analysis* by Brian D. Warner. Indeed, whichever method you use it would be wise to have this book on hand.

Equipment Needed

To do asteroid lightcurve photometry you need the same tools of the trade as described in Chap. 10 – a telescope, a polar-aligned mount, a CCD camera, a PC, and the ability to obtain calibration frames, e.g., dark frames and flat fields. Calibration frames are of utmost importance as, without them, accurate measurements of magnitude cannot be made. The basics of CCD imaging and the reasons why calibration frames need to be applied was explained in Chap. 10.

Images of the author's light bin for obtaining flat fields and light-tight cap for obtaining dark frames also appear in Chap. 10, but more detailed diagrams are shown in Figs. 12.6 and 12.7. There are other ways of obtaining flat fields, imaging

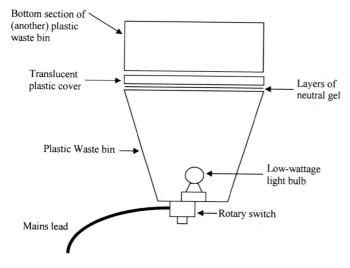

Fig. 12.6. Light bin (Diagram by the author).

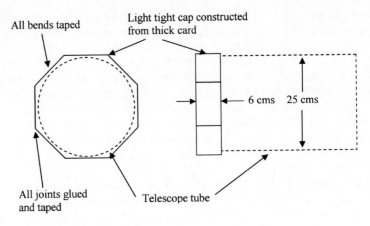

All bends taped

Light tight cap constructed from thick card

6 cms   25 cms

All joints glued and taped

Telescope tube

Fig. 12.7. Light-tight cap (Diagram by the author).

the sky at twilight, for example, the objective being to produce a low-level uniform light source. Some CCD cameras are fitted with shutters, so a dark frame can be obtained with the camera shutter closed, but dark frames and flat fields should always be obtained when it is dark to avoid any stray light creeping into the camera.

Differential Photometry

Choosing a Target

Here are some guidelines to follow if you want to avoid wasting time 'chasing' unsuitable targets and determining magnitudes, which may have a considerable uncertainty:

- Choose asteroids higher than 25–30° altitude so that you stay well clear of the murk adjacent to the horizon. (Also the light from low-altitude objects has further to travel through Earth's atmosphere and magnitude and thus signal will be decreased, signal-to-noise ratio (SNR) being all important.)
- Choose an asteroid that is in the E or SE at the start of the imaging session, as it can be imaged for a longer period of time than one that is close to or has already passed the meridian.
- Avoid asteroids at or close to opposition, say within 10°, as the 'opposition effect' (see Chap. 14) can distort the lightcurve.
- Do not attempt to image objects or use comparison stars that are too faint to ensure a sufficiently high SNR, ideally >20 (in my case using a 0.25 m telescope this means working on objects brighter than approximately magnitude 14). Limiting magnitudes for various telescopes are typically: 0.2 m/13.5, 0.4 m/15.0, 1.0 m/17.0 and 2.0 m/18.5. The stars in the FOV can be checked for magnitude and known variability using planetarium programs such as *Megastar* or *Guide*. The SNR can be improved by stacking multiple images and taking short exposures, which minimizes trailing due to poor tracking and telescope vibration.

My Newtonian reflector is particularly susceptible to even light winds, so I tend to limit exposure times to around 30 s.

- Choose an exposure time that avoids excessive trailing (say by more than 2 or 3 pixels) due to the motion of the asteroid (exposure time in seconds = 60 × (image scale in arcsecs/pixel)/(motion in arcsecs/min) as previously mentioned in Chap. 11).
- To prove to yourself that you can achieve reasonable results pick an asteroid that has a well-defined period of rotation and an amplitude of 0.2 magnitudes or greater. Even better choose one that has a period such that you can obtain a complete lightcurve in a single night's imaging, so that you won't have the added complication of combining several nights' work. The relevant data can be obtained by accessing the Lightcurve Parameters Page on the Collaborative Asteroid Lightcurve Link (CALL) website. However do be aware that the amplitude of the lightcurve differs from opposition to opposition and will not necessarily be the same as the published value.

As an example we will work through the method the author used to obtain a lightcurve of asteroid (423) Diotima on the night of January 12/13, 2005. This was chosen from the list of asteroids on the Association of Lunar and Planetary Observers' Photometry and Shape Modeling web page. Its rotational properties suggest it belongs to a class of large lightcurve-amplitude, rapidly rotating asteroids found most commonly among those in the size range 100–300 km in diameter. This asteroid was well placed for a lengthy observation (high in the southeast) and of a magnitude (12.1) which would give a high signal-to-noise ratio and thus minimize errors. The short predicted period (4.8 h) meant that I should be able to observe a complete rotation in a single night, weather permitting.

My observing log also records that the *Deep Impact* spacecraft was launched that same day, beginning its journey to comet Tempel 1, and I also noted that I viewed comet C/2004 Q2 Machholz through 10×50 binoculars. So if you don't want to retreat to your warm and comfortable study then there are other objects you can observe. Why not count meteors, hunt for comets or novae, or watch out for the International Space Station to pass by?

Imaging

I followed my usual observatory set-up procedure:

- Switch on laptop PC, GPS receiver, CCD camera, and CD writer.
- Polar and three-point align telescope.
- Load the latest orbital elements into *Megastar* and plot the track of the asteroid.
- Locate target visually and then replace the CCD framing eyepiece with the CCD camera.

Test images were taken to check focus and ensure that the maximum pixel intensity was of the order of 50%, to ensure that the CCD camera was working over the linear portion of its response curve – verified with the facility in the *Astroart* software package, which controls imaging. Initially I set the exposure time to 30 s but subsequently reduced this to 20 s to prevent overexposure. If you want to maximize your imaging time then you can complete all your set-up as soon as the sky is dark enough to see the necessary stars to carry out polar- and three-point alignment (or whatever you need to do prior to imaging). Another advantage of doing this work early is that it is still light enough to find your way around without the need for artificial light.

Imaging commenced at 19:35 UT and ended at 02:16 UT the following morning. I tend to take images in batches of 60, so that I am prompted to check that all is going according to plan every once in a while. So the night went as follows:

- 'A' sequence: 60 × 30 s exposures with a 30-s delay between each (enables the PC clock to reset, although nowadays I usually use just a 10-s or 15-s delay).
- 'B' sequence: 6 × 30 + 53 × 20 s exposures with a 30-s delay between each.
- 'C' sequence: 46 × 20 s exposures with a 30-s delay between. Sequence abbreviated as telescope needed to be reversed.
- 'D' sequence as for 'B.'
- 'E' sequence as for 'B.'
- 'F' sequence: 51 × 20 s exposures with a 30-s delay between each, giving a grand total of 337 unfiltered images. Imaging terminated for the night, as the sky had clouded over.
- Calibration frames obtained: 5 × 30 s dark frames, 5 × 20 s dark frames, 5 × 3 s flat-dark frames, and 5 × 3 s flat fields. It is recommended that at least five of each type of calibration frame be obtained and further sets obtained if the air temperature varies by more than ±5°F (2°C).

All images, including calibration frames, were saved in FITS format.

Image Processing and Analysis Using *MPO Canopus*

Images were processed the following day using the *Minor Planet Observer's (MPO) Canopus* software. *Canopus* is the software of choice for many if not most amateur astronomers for generating asteroid lightcurves. With this software you can calibrate images, measure them, generate a lightcurve, and integrate photometry obtained on different occasions by the same or multiple observers, although the last option was not required on this occasion as all the required images were obtained during a single imaging session. All the facilities are fully explained in the lessons, manuals, and help provided with the software, but the relevant operations to obtain a lightcurve are summarized here.

Setting the Configuration. Select 'File/Configuration' and check that the configuration is correct for your circumstances. Various configurations can be set up by entering a new name in the 'Profile' box and entering the relevant data. The 'General' tab is self-explanatory. Under 'Photometry/Miscellaneous' select 'Instrumental' in the 'Photometry Magnitude/Method' box and uncheck 'Heliocentric times,' this facility being more relevant for variable stars than asteroid lightcurves. In 'Photometry/Plotting Options' select 'Center on mean mag.' The other tabs are largely irrelevant for differential photometry.

Creating a Session. A session is a contiguous set of observations that use the same set of comparison stars (for differential photometry). Open a session by selecting 'Photometry/session' and enter the data relevant to your observations, e.g., object, mid date/time, telescope, focal length, CCD camera, temperature, and exposure time. Click on the 'Calc/M/D/P' button and select the asteroid (in this case 423) from the pop-up 'Asteroid look-up' window. This inserts values in the 'E(stimated) Mag,' 'E(arth) Dist,' 'S(un) Dist,' 'RA,' and 'Dec' boxes. So that observations made over several days can be combined *Canopus* calculates the reduced magnitude from your measurements of apparent magnitude (which will be explained in Chap. 14). Select 'Save' then 'OK' to save the session information.

E Mag, E Dist, S Dist, RA, and Dec are calculated from orbital elements downloaded from the Lowell Observatory (Asteroid Orbital Elements – ASTORB) or Minor Planet Center (MPC Orbit – MPCORB) databases. It is therefore important that these databases are kept current by downloading the latest orbital elements. This can be done from the 'Pages/Conversions' window.

Calibrating Images. Master dark and master flat images are created using the Batch process tool, which is accessed by clicking on 'Utilities/Image Processing/ Batch process':

- Create a master dark frame by median combining dark frames.
- Create a master flat-dark by median combining flat-dark frames.
- Merge the master flat-dark with each flat field to create sub-master flats.
- Median combine the sub-master flats to create a master flat field.

To calibrate images (again using the Batch process tool):

- Select 'Action – Merge Dark and Flat.'
- Select the required master dark and master flat frames.
- Select images to be calibrated.
- Click on Process and image names will be displayed in the lower right hand window as they are processed.

Setting Apertures. The three boxes in the center of the toolbar are used to set the measuring or star aperture, dead zone (difference between star aperture and inner sky annulus), and sky annulus (outer minus inner radii), respectively (see Fig. 12.8).

Generating a Light Curve. If you are unsure as to the position of the asteroid use the Blinker – select Pages/Blinker. It is worthwhile checking that the 'gunsight' is in the correct position on all images to be blinked. It can move from its original spot if further images are opened.

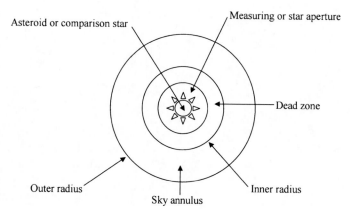

Fig. 12.8 Apertures (Diagram by the author)

The first part of the procedure is to identify the asteroid or target and the comparison stars to be used (see Fig. 12.9). Select the relevant session by clicking on 'Photometry/Session,' highlighting the relevant session and then clicking on 'OK.' Use the Lightcurve Photometry Wizard to guide you through the process of measuring your images:

- Select 'Photometry/Lightcurve Wizard' and then the first image of the sequence in the pop-up window. That image will then be displayed. Those with equatorial mounts will be pleased to know that *Canopus* copes quite easily with telescope reversals. Just check the 'Measure two sets' box in the Lightcurve Photometry Wizard window before opening the first image of the first (prior to' scope reversal) set. Note that the same sequence of comparison stars must be used in both sets.
- Identify the asteroid (target) and comparison stars.
- Open the last image in the sequence and identify the same target and comparison stars.
- Open the first image obtained after telescope reversal and identify the same comparison stars as those identified prior to telescope reversal.
- Do the same with the last image obtained after telescope reversal.

Having identified the asteroid and comparison stars the next part of the procedure is to measure the (instrumental) magnitudes of those objects:

- Select all the images you wish to use from the Photometry Image List, which is displayed on completion of the previous step. This opens the Images window.
- Double clicking on the first of the images listed displays that image with the measuring apertures located over the asteroid and comparison stars.
- If the image is acceptable (i.e., not trailed or maybe dimmed by a passing cloud), then click on the 'Accept' button. *Canopus* allows you to accept or reject images and adjust the positions of the annuli for each image. Good tracking will

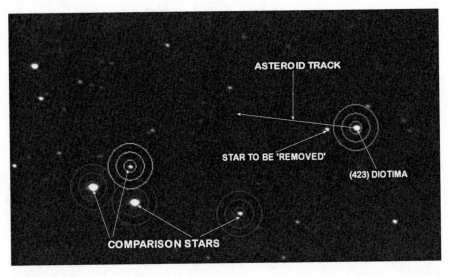

Fig. 12.9. (423) Diotima and comparison stars (Credit: *MPO Canopus*).

necessitate fewer adjustments of the positions of the annuli, as the positions of the stars and asteroid will vary little from image to image, and of course trailing of the star and asteroid images will be minimal. *Canopus* includes a feature, StarBGone!, for use with asteroid targets that move close to stars during an imaging run. In simple terms this removes selected stars from an image just before it is measured, thus avoiding the loss of data.

– When all acceptable images have been measured close the Images window.

The instrumental magnitudes will be listed in the object and comparison instrumental magnitude columns under the 'Observations' tab in the relevant session.

Finding the Lightcurve Period. Finding the period is not necessarily easy or automatic. It does help to have a rough idea of what it might be and to have observations that span a number of complete cycles. It is reasonable to assume that the lightcurve has a typical double-peaked shape as mentioned earlier in this chapter. To display your data graphically and find the period of the lightcurve and hence the rotational period of the asteroid:

– Select the required session – 'Photometry/Session/OK.'
– Bring up the Photometry page – 'Pages/Photometry.'
– First display your data (see Fig. 12.10) for the whole of the period or periods of observation by checking the 'Raw' box.
– The Results window, displayed prior to the chart, will highlight the best-fit period. (But this should not necessarily be taken for granted, as to quote from one of the *Canopus* lessons, 'it cannot be stressed strongly enough that finding the period of a lightcurve is just as much art, experience, and even luck as it is science.')

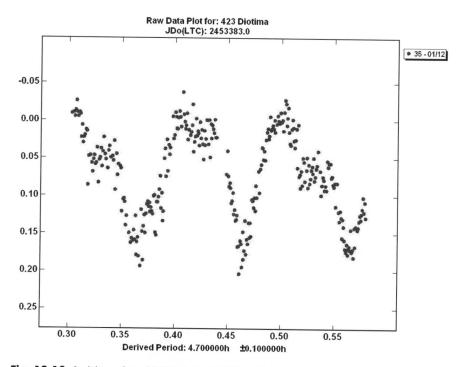

Fig. 12.10. Raw lightcurve of asteroid (423) Diotima (Credit: *MPO Canopus*).

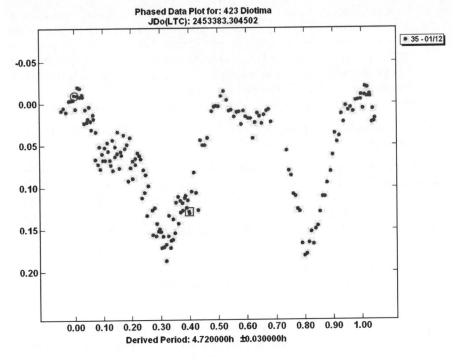

Fig. 12.11. Phased lightcurve of asteroid (423) Diotima (Credit: *MPO Canopus*).

- In the case of (423) Diotima the suggested period was 4.7 h, and it can be seen that the lightcurve in Fig. 12.10 does appear to repeat itself in that time – the observations being made over a 6 h 41 min period.
- Experimenting with the settings for (Min)imum period, Size, and Steps, and with Raw unchecked gave a solution of 4.74 ± 0.03 h (see Fig. 12.11). This plot has been cleaned up by removing two obviously rogue measurements and binning adjacent points. The 'Bin' box allows you to select the number of measurements to bin, and 'Max. Diff,' as its name suggests, is the maximum time in minutes between any two consecutive measurements.

The lightcurve deviates from a smooth double-peaked shape at 0.1 and 0.6 of a period. A paper published by L. G. Karachkina and V. V. Prokof'eva in 2003 concluded that (423) Diotima was a binary asteroid and one of a family of 411 members in a 9/4 resonance with Jupiter (refer to Chap. 5 for more on resonances).

A New Approach

Slow Rotators

Asteroid (1909) Alekhin is at the other end of the spectrum as far as rotational periods are concerned, as will become clear in this example of obtaining a lightcurve using a robotic telescope and the methodology mentioned previously and fully described in Appendix C of this book. In the spring of 2009 the author identified this asteroid as a suitable target from a list published by the Magnitude Alert Program (MAP) (but more on that in Chap. 14). A search of several websites that list rotational periods of asteroids – Geneva Observatory (Asteroid and Comet rotation curves), the Lightcurve Parameters Page on the CALL website, the Ondrejov Asteroid Photometry Project dataset, and the Standard Asteroid Photometric Catalog – all drew a blank, so perhaps this is the first time that the rotational period of this asteroid has been determined. The cause of such slow rotation is still a mystery but may be due to the Yarkovsky effect described in Chap. 5.

Before the advent of the method described here there were a number of difficulties in obtaining lightcurves of very slow-rotating asteroids (and this may be the reason few have been made). Many nights' observations are required, and, during this time, the asteroid will move through many star fields, requiring the use of different comparison stars. To properly combine the segments of the lightcurve requires that the actual magnitude of the asteroid be measured as opposed to the differential magnitude, and thus the observer would have to resort to the more complex all-sky or absolute photometry. The study of slow-rotating asteroids is thus an ideal activity for amateur astronomers, as the amount of telescope time we have is less limited, compared with professional astronomers, and we now have a simple method of ascertaining magnitude even when the object being measured is moving through several star fields.

Imaging

The path of asteroid (01909) Alekhin showing the CCD FOVs for the nights of observation is shown in Fig. 12.12. As is obvious from the figure different comparison stars have to be used, as the asteroid tracks from west to east.

The Sierra Stars Observatory Network (SSON) robotic telescope located in California was used to obtain the images. This is a very easy system to use, as (having previously purchased the necessary time) one simply logs in and selects:

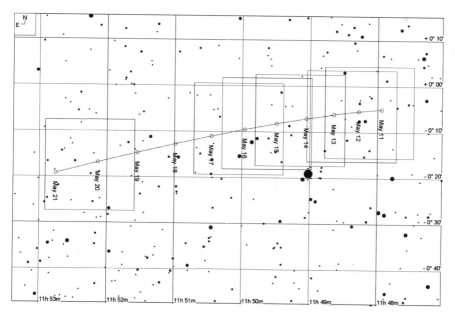

Fig. 12.12. Track of (01909) Alekhin, including CCD frame overlays (Credit: *Megastar*).

- Object to be imaged from the Moving Object Catalog, (01909) Alekhin.
- Date and time if required, but not specified in this case.
- Filter, V.
- Exposure time, 60 s.
- Number of images to be taken (calibration frames will be automatically applied).
- Time between images, varied but typically 60 min.

Having set up the job you sit back and wait for notification by e-mail that the job has run and your images are available for downloading. If bad weather or technical problems prevent any images from being taken they are automatically rerun the next night (if you have specified a date and time the job is canceled and you have to resubmit it).

Images were obtained on 17 nights between March 28 and May 27, 2009. Figure 12.13 is part of a stack of images obtained on May 12, 2009, showing the asteroid at hourly intervals and its direction of movement. Using a robotic telescope in a different part of the world presents its own difficulties, e.g., what part of the sky is visible; what is the local time compared with UT; what is the local weather forecast; what are the hours of darkness; at what times does the target rise, culminate, and set; and what is its altitude. Purchasing a planisphere for the latitude in question solved the first problem and data on the SSON website answered the others. Planetarium programs such as *Megastar* can be set up to reflect different locations and time zones, which assists in scheduling imaging.

Fig. 12.13. Stack of images of asteroid (01909) Alekhin (Credit: *Astrometrica*/SSON).

Image Processing and Analysis

As mentioned earlier the SSON images are automatically calibrated. The magnitude of the asteroid was measured using *Astrometrica*, the methodology described in Appendix C of this book, and the data formatted to allow it to be imported into the sessions, one for each night's imaging and previously set up in *Canopus*. The import format is merely date, UT, and magnitude (see Table 12.2). The increase in brightness of the asteroid in the space of the 4 h is quite noticeable (remembering that a *decrease* in the magnitude figure represents an *increase* in the brightness of the object).

The only *Canopus* configuration setting that needs be changed from those described previously is to set 'Photometry Magnitudes' to 'Transformed' rather than 'Instrumental.' Initially *Canopus* suggested a period of approximately 74 h, but the lightcurve was displayed as a single peak, which is extremely unlikely. Forcing a period of twice that duration produced the more typical double-peaked lightcurve, and as more observations were made *Canopus* also 'decided' that this was a sensible solution. A little manual adjustment produced the lightcurve shown in Fig. 12.14.

Table 12.2. Data formatted for input into *Canopus*

Date	UT	Magnitude
2009/04/29	04:17:50	15.52
2009/04/29	05:17:50	15.51
2009/04/29	06:17:30	15.49
2009/04/29	07:17:30	15.46
2009/04/29	08:32:30	15.43

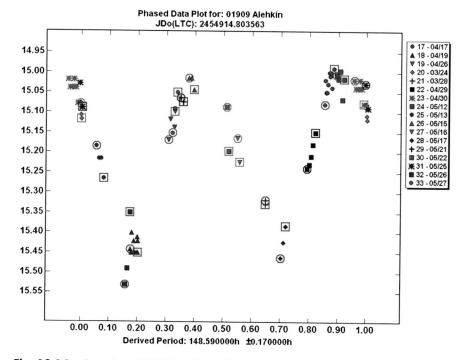

Fig. 12.14. Lightcurve of asteroid (01909) Alekhin (Credit: *MPO Canopus*).

A combination of some bad weather, the low altitude of the asteroid, shorter nights towards the end of May, and the waxing Moon prevented further observations, and hence the gaps in the lightcurve.

Whichever photometry software package you use you will need to set the configuration to suit your circumstances. Once you have done this it is advisable to print out a copy for future reference. To help ensure the validity of any future measurements it is worthwhile saving and subsequently checking a few what might be called 'gold standard' images. These are ones where you have previously measured the magnitude of an asteroid and are sure that it is correct. If you get different results then you have a problem that needs resolving before further measurements of new images are made.

Conclusion

Some will have you believe that photometry is something of a black art, and if you are not achieving an accuracy of ±0.001 magnitude it isn't worth doing. Not so! If you can achieve ±0.05 then even professional astronomers will be more than happy to receive your results, as will be described in the next chapter.

You just have to be methodical (or pedantic as some describe it). Work out a procedure, write it down, and follow it rigidly.

Chapter 13

Lightcurve Photometry Projects

Having worked your way through the previous chapter you will now know how to obtain images suitable for photometry, calibrate them, and construct a lightcurve. This chapter will direct you to lists of asteroids suitable for lightcurve work and describe ongoing projects to which you can contribute. To quote from a paper to which a number of amateurs contributed, "A substantial part of the photometric data was observed by amateur astronomers. We emphasize the importance of a coordinated network of observers that will be of extreme importance for future all-sky asteroid photometric surveys."

Although the various sources mentioned will list Right Ascension and Declination for the asteroids in question you should always obtain the latest orbital elements and/or an ephemeris for your location from the Minor Planet Center. Once you feel competent in this work please do publish your results. The source of information will usually indicate where results are to be sent. The chapter concludes with two 'extreme' examples of what amateur astronomers can achieve and also gives dwarf planets a mention.

Those sources or projects that include the brighter asteroids are described first, followed by those with more challenging targets. A guide to the size of telescope required to obtain accurate lightcurves for asteroids of various magnitudes was given in the previous chapter. Please note that the projects listed here were active at the time this book was written. Unfortunately some may have been terminated (and new ones implemented, to look on the positive side) by the time you read this, but they can still give you a very good idea of the kind of work you can carry out.

The *Handbook of the British Astronomical Association*

Lightcurve targets are listed in the above publication, which can be obtained from the BAA, and on its Asteroids and Remote Planets Section website. Listed are asteroids, magnitude 14 or brighter, for which either no rotation period has been determined or the period is based on partial lightcurves and may be incorrect – Table 13.1 provides an example of the asteroids included. At least with these you have a starting

R. Dymock, *Asteroids and Dwarf Planets and How to Observe Them*,
Astronomers' Observing Guides, DOI 10.1007/978-1-4419-6439-7_13,
© Springer Science+Business Media, LLC 2010

Table 13.1. Example lightcurve opportunities in 2009 from the *Handbook of the British Astronomical Association*

Asteroid number	Asteroid name	Date when brightest m d	V Mag	Δ (AU)	Dec.	Elong. (°)	U	Period (h)	Amplitude (Mag)
29943	1999 JZ78	1 01.3	13.9	1.676	+29	174	–		
786	Bredichina	1 07.4	13.3	2.347	+28	174	–		
946	Poesia	1 15.9	13.8	1.695	+23	178	–		
136849	1998 CS1	1 15.9	12.4	0.035	+29	128	–		
366	Vincentina	1 19.4	13.3	2.344	+31	169	1	15.5	0.08
168	Sibylla	1 23.1	12.8	2.457	+13	174	1	23.8	>0.3
271	Penthesilea	1 31.2	13.9	2.029	+19	178	–		
566	Stereoskopia	2 03.0	13.0	2.554	+22	174	2	17	0.08
850	Altona	2 06.5	14.0	2.133	+22	174	1	4	
500	Selinur	2 11.1	13.3	1.888	+6	172	1	>8	>0.05
602	Marianna	2 12.8	13.0	2.415	+16	177	1	30	0.3
	1999 AQ10	2 17.1	13.1	0.014	−24	119	–		

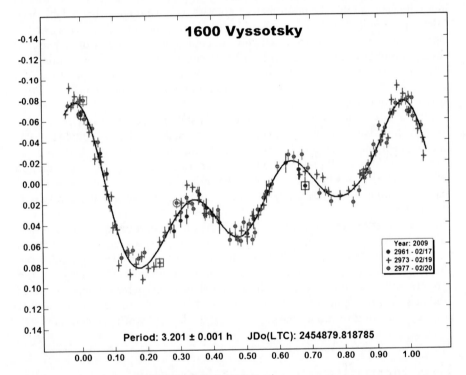

Fig. 13.1. Hungaria family asteroid (1600) Vyssotsky (Credit: Brian D. Warner).

point, and there are several with suspected periods of just a few hours where you could obtain a complete lightcurve in one night.

It is more than likely that a lightcurve will be of the typical double-peaked variety, but there are exceptions, due to the asteroid being a binary, having a satellite, or tumbling, i.e., rotating, about more than one axis. Figure 13.1 is an example of a triple-peaked lightcurve. To be certain of the period you should try to observe the asteroid over at least three complete cycles. These asteroids are suitable for both

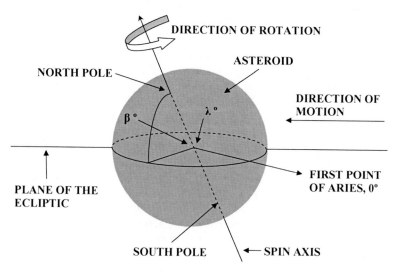

Fig. 13.2. Asteroid pole position (Diagram by the author).

differential and absolute or all-sky photometry, but the former is of course easier. The new approach described in Chap. 12 can also be applied, providing you use a V filter or have a CCD camera with a very similar response.

The reliability code, U, which features in many of the lists, is an indication of the accuracy of the indicated period:

1. Result based on fragmentary lightcurve(s) may be completely wrong.
2. Result based on less than full coverage, so that the period may be wrong by 30% or so. Also used to indicate cases where an ambiguity exists as to the number of extrema between lightcurves. Hence the result may be wrong by an integer ratio.
3. Secure result with no ambiguity, full lightcurve coverage.
4. In addition to full coverage, denotes that a pole position is reported.

The pole position (see Fig. 13.2) is defined in terms of ecliptic coordinates measured westwards from the First Point of Aries or Vernal Equinox (λ degrees) and northwards (positive angle) or southwards (negative angle) from the plane of the ecliptic (β degrees). The north pole is defined in a similar way to Earth's North Pole in that, if one were to look down on it, the asteroid would appear to be rotating counterclockwise. Thus in Fig. 13.2 the pole position might be defined as (approximately) ecliptic longitude (λ), 270° and ecliptic latitude (β), +60°.

Minor Planet Bulletin (MPB) Projects

The MPB is the official publication of the Minor Planets Section of the Association of Lunar and Planetary Observers (ALPO). The MPB is available free on-line (contributions are welcome) and in print (by subscription) and is regarded by many amateur astronomers as *the* journal to which they should submit their results for publication. Each quarterly issue lists targets of opportunity, explained in the following paragraphs, for the next 3 months.

Shape and Spin Axis Modeling

An asteroid can be described as a triaxial (or scalene) ellipsoid (see Fig. 13.3), where a, b, and c are the lengths of the three axes with a≥b≥c. In this diagram view B is that seen from the direction of X, and view C from Y. It is possible to calculate the ratios of these axes from a lightcurve, but the mathematics is a little too complicated to discuss here. Asteroids will usually rotate around one of their shorter (or minor) axes, being somewhat unstable if rotating around the long (or major) axis. The larger, differentiated asteroids and dwarf planets have a lower oblateness (are more spherical), measured by a–c in Fig. 13.3, than smaller, solid bodies.

The *MPB* lists those asteroids needing only a small number of lightcurves to allow shape and spin axis to be modeled. It can be seen from Table 13.2 that the rotational periods and lightcurve amplitudes are known and the asteroids are quite bright, thus making them relatively easy targets.

For modeling work, absolute photometry is recommended, e.g., absolute values of magnitude put onto a standard system such as Johnson V. If this is not possible good differential photometry is just as acceptable. When working any asteroid, keep in mind that the best results for shape and spin axis modeling come when lightcurves are obtained over a large range of phase angles within an apparition.

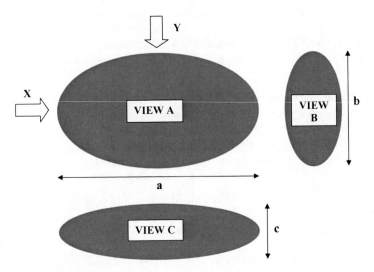

Fig. 13.3. A triaxial ellipsoid (Diagram by the author).

Table 13.2. Example of shape-modeling targets

Number	Name	Brightest Date	Mag	Dec	U	LCDB Data Period	Amp
1622	Chacornac	4 22.7	13.8	−18			
2610	Tuva	4 25.8	14.7	−13			
207	Hedda	5 01.6	12.1	−17	1	>12	0.03
1197	Rhodesia	5 02.4	12.9	−28	2	16.06	0.22–0.32
93768	2000 WN22	5 03.8	14.7	8			
11200	1999 CV121	5 08.9	14.4	−15			
2639	Planman	5 09.8	14.8	−07			
5773	1989 NO	5 10.3	14.9	−25			

If at all possible try to get lightcurves not only close to opposition, but before and after, i.e., when the phase angle is 15° or more. For those advanced amateur astronomers who want to produce their own shape models there is software, *LCInvert*, available from the Minor Planet Observer. It must be pointed out however that data from more than one apparition is required to model the shape and spin axis of an asteroid.

Determination of Rotational Period

Included are those asteroids reaching a favorable apparition (one during which the asteroid is brighter than usual), exceeding magnitude 15 at brightest, and which have either no or poorly defined lightcurve parameters. Examples are shown in Table 13.3. Observers are encouraged to seek help from others to achieve the goal of determining the rotation rates of these asteroids. All three methods described in Chap. 12 are applicable to this work, but differential photometry is the norm.

In Support of Planned Radar Targets

The world's two primary radar facilities used for asteroid observation are the Arecibo Observatory in Puerto Rico and the Goldstone Solar System Radar (see Fig. 13.4),

Table 13.3. Example of lightcurve targets from the *Minor Planet Bulletin*

Number	Name	Brightest Date	Mag	Dec	Period (h)	Amp	U
146	Lucina	4 10.3	11.8	+10	18.557	0.08	3
54	Alexandra	4 16.3	11.3	−26	7.024	0.10–0.31	3
14	Irene	4 20.7	8.8	+01	15.06	0.08–0.12	4
451	Patientia	5 04.8	11.4	−01	9.727	0.05–0.10	3
409	Aspasia	5 06.1	10.4	−22.9	9.022	0.10–0.14	3

Fig. 13.4. Goldstone 70 m (233 ft) antenna (Credit: NASA/JPL Caltech).

both in the United States. Arecibo can 'see' further than Goldstone, but Goldstone's steerable dish can track an asteroid for longer. Originally Arecibo could only target asteroids as distant as the inner edge of the Main Belt but can now observe asteroids across the whole width of that region.

The first radar observation of an asteroid, (1566) Icarus, was made in 1968 and by June 30, 2009, 365 NEOs and Main Belt asteroids had been targeted. Figure 13.5 shows a radar image of asteroid (4179) Toutatis made during its close approach to Earth in December 1992, when it was approximately 4 million km (2.5 million miles) from Earth. Toutatis appears to be a contact binary – two bodies orbiting one another and actually touching. The observations were made using the Goldstone 70 m (230 ft) dish by a team led by the late Dr. Steve Ostro.

Radar is a powerful source of information about the physical properties and orbits of asteroids. It can measure the distance to an asteroid and its radial velocity; from such measurements a detailed three-dimensional model can be created and its rotation rate precisely defined. In addition the radar echo can reveal details of the surface roughness of an asteroid, being able to differentiate between a very smooth surface and a rocky one or one covered in boulders. Radar observations also allow orbits to be very accurately defined, thus allowing the passage of NEOs to be precisely predicted as they approach Earth.

Optical (CCD) observations made at the same time as radar observations can be helpful. Accurate last-minute astrometry ensures the dish is pointed in the right direction, and lightcurve photometry complements the rotation rate and shape determined by the radar observations. Radar observations have led to the discovery,

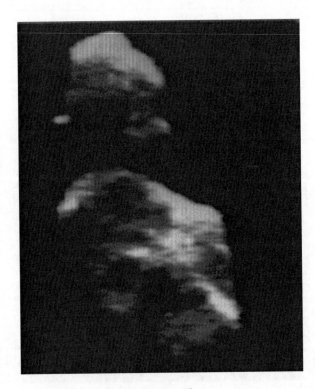

Fig. 13.5. Radar image of asteroid (4179) Toutatis (Credit: Steve Ostro, JPL).

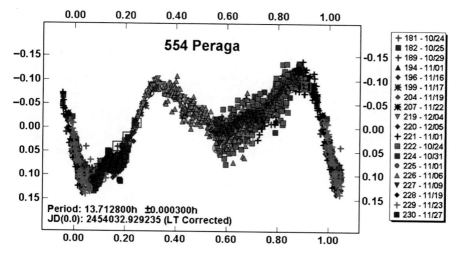

Fig. 13.6. Lightcurve of (554) Peraga (Credit: Roger Dymock, Martin Crow, David Higgins).

or confirmed the existence, of several binary asteroids. Differential photometry is satisfactory for this type of work but high precision, 0.01–0.03 magnitudes, is preferable, although 0.05 or better is usually quite acceptable. A list of potential radar targets for which optical observations may be requested is maintained by Lance A. M. Benner. Scheduled Arecibo and Goldstone radar observations including whether or not astrometry and lightcurve data are required can be found on their respective websites.

Specific requests for such observations are also occasionally published on the Minor Planet mailing list. The result of one such request by Dr. Ellen Howell, who was using the Arecibo observatory, can be seen in Fig. 13.6 – a lightcurve of asteroid (554) Peraga constructed by Australian amateur David Higgins from observations made by himself, Martin Crow, and this author. Prior to this work the period of (554) Peraga was estimated to be 13.62 h, so it was comforting when the estimate by Martin Crow and myself produced a different value of 13.709 ± 0.002 h, which was confirmed by David Higgins' estimate of 13.714 ± 0.002 h – a difference of just 18 s!

Collaborative Asteroid Lightcurve Link (CALL)

CALL is run by Brian D. Warner, and its website has links to various projects that are summarized below.

Mikko Kaasalainen's Shape Modeling Program

It was mentioned at the beginning of this chapter that this is an area of astronomy where professionals and amateurs regularly work together. As an example

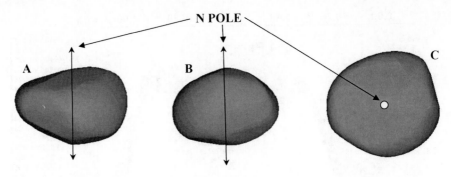

Fig. 13.7. Shape model of asteroid (423) Diotima (Credit: Mikko Kaasalainen).

this author imaged asteroid (423) Diotima and forwarded the results to Mikko Kaasalainen at the University of Finland in Helsinki as part of Kaasalainen's shape-modeling program. The lightcurve (see Fig. 12.10) was subsequently published in the journal *Astronomy and Astrophysics* in a paper entitled 'Physical models of ten asteroids from an observers' collaboration network.' The shape model, which those observations helped to derive, is shown in Fig. 13.7. A and B are equatorial views 90° apart, and view C is looking down on the north pole of the asteroid.

A list of asteroids, both NEOs and Main Belt, requiring further observations as part of this program can be accessed via the CALL website. Those interested in shape modeling might like to access the Database of Asteroid Models from Inversion Techniques (DAMIT) at the Astronomical Institute of the Charles University, Czech Republic. Results and the original data for a large number of asteroid models are available at this location, and a number of papers on the subject are available on Mikko Kaasalainen's website.

Koronis Family Asteroids Rotation Lightcurve Observing Program

Dr. Stephen M. Slivan, of the Massachusetts Institute of Technology (MIT), is studying the spin properties of members of the Koronis family of Main Belt asteroids, which are thought to be the shattered remains of a single body destroyed in a collision. Rotational periods of asteroids in this family average between approximately 6 and 18 h. The spin axes of a number of the largest asteroids in this family are, unusually, aligned in obliquity (the angle between the 'equator' of the asteroid and the ecliptic). Further lightcurves are needed to determine the spin axis of more members of this family. The website associated with this program includes an 'observing targets calculator,' which allows targets to be selected depending, for example, on an observer's location, date, time, magnitude range, and observing period. Many if not most of the objects listed are typically magnitude 15 or fainter, so a large amateur telescope with a primary mirror diameter of at least 0.4 m (16 in.) is necessary to achieve a satisfactory signal-to-noise ratio. Stacking images may enable a slightly smaller telescope to be used.

Jupiter Trojans

Joshua Emery of the University of Tennessee is a member of a team that is using the Spitzer Space Telescope to observe Jupiter Trojan asteroids (Spitzer observes in the infrared between 3 and 180 µm). The objects on their target list have no or poor lightcurve data; thus, such data would make analysis of the Spitzer data more robust, especially if the lightcurve observations are close in time to the Spitzer observations. Spitzer only announces its schedule a few weeks in advance, so potential participants should frequently review the target list, which can be accessed via the CALL website. The data that is required is lightcurve amplitude, rotational period, and absolute magnitude, and the closer to the time of the Spitzer observations that these are obtained so much the better. Typically target asteroids are around magnitude 15 and 16, requiring at least an 0.4 m (16 in.) telescope. Ideally measurements should be made to a precision of 0.01–0.03 magnitudes, but slightly less precise data should be acceptable providing that the precision is quoted.

Karin Family Asteroids

The Karin family is a small group of asteroids, with (832) Karin being the primary body. This family, discovered by David Nesvorny and William F. Bottke, is believed to have been created approximately 5–6 million years ago by the collision of another asteroid with the group's 15 km diameter progenitor, making it one of the youngest families known. The aim of the project is to study the early stages of an asteroid family's evolution. Of particular interest are lightcurves to determine rotation rates and identify if any of the family are tumbling. These curves can be combined with those obtained in other years to see what effect, if any, YORP (described in Chap. 4) has had on the family members.

A list of targets for a specific location can be generated from the project's website. Most of the asteroids for which lightcurves are required are magnitude 19 or 20, which are beyond the reach of most amateurs, although access to a large robotic telescope could allow satisfactory measurements to be made.

Ondrejov Asteroid Photometry Project

The goal of this project is to discover asynchronous binary asteroids among the populations of small diameter (<10 km) near-Earth asteroids, Mars crossers, and inner Main Belt asteroids. Asynchronous binaries are those where the rotational period of the secondary or moon differs from its orbital period around the primary. This differs from our own Moon, which is locked in synchronous rotation around Earth – its period of rotation being equal to its orbital period and hence we always see (give or take a few percent) the same side. A lightcurve that differs from the 'standard' double-peaked variety may be an indication that the asteroid has a satellite. Such deviations from the norm, or events, as they are known, will cause the lightcurve to vary from cycle to cycle and allow the rotational period of the primary to be distinguished from the orbital period of the secondary.

Figures 13.8 and 13.9 illustrate the lightcurve of (6265) 1985 TW_3 – the rotational period of the primary body being 2.71 h and the orbital period of the secondary

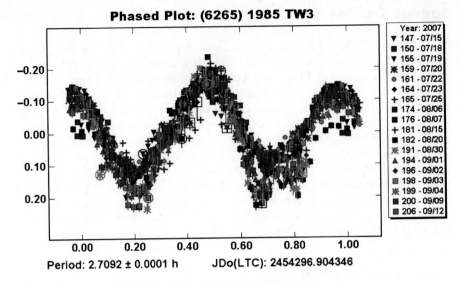

Fig. 13.8. Lightcurve of binary asteroid (6265) 1985 TW₃ (Credit: David Higgins).

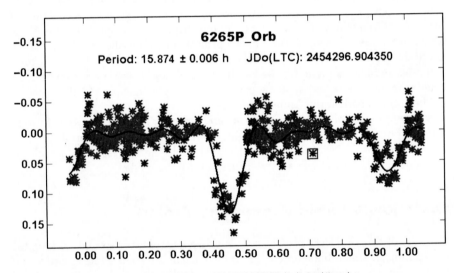

Fig. 13.9. Lightcurve showing eclipsing events of binary asteroid (6265) 1985 TW₃ (Credit: David Higgins).

being 15.87 h. The lightcurve (see Fig. 13.8) of the binary system is made up of three components due to the rotation of the primary body, the rotation of the secondary, and the orbit of the secondary around the primary, which will include eclipsing events. Figure 13.9, which was constructed by subtracting the median values from the data used in Fig. 13.8, shows the combined effects of the rotation of the secondary and the orbit of the secondary around the primary. Assuming that most secondaries are tidally locked to their primaries, the rotational period of the secondary will have the same value as its orbital period, and thus Fig. 13.8 can be treated as representing just the orbital period. Readers interested in this project might like to contact Petr Pravec of the Ondrejov Asteroid Group, whose e-mail address can be found on the project's website.

'Extreme' Lightcurves

By now you may feel you have seen enough lightcurves to last a lifetime, but here are two more, one of a very distant dwarf planet and one involving a very fast rotating asteroid, hence the description 'extreme.'

In 2007 UK amateur astronomer John Saxton took 532×30-s unfiltered images of dwarf planet 2003 EL_{61} (now known as (136108) Haumea) using his 30 cm (12 in.) Meade SCT and a Starlight Xpress MX916 CCD camera. At the time 17.4 magnitude Haumea was over 50 AU from Earth – a 'challenging target,' as John put it. Using his own *Lymm* software he constructed the lightcurve shown in Fig. 13.10 and estimated the spin period to be 3.9 h, which agrees with estimates made by professional astronomers. ('Flux' on the vertical axis is a measure of brightness on a linear scale in relation to a comparison star – whereas normally 'magnitude' is quoted.)

Unlike asteroids, dwarf planets are by definition usually spherical in shape, and their lightcurves, dictated by surface features and albedo rather than shape, will not necessarily exhibit the 'normal' double-peak features. However Haumea, being football or rugby ball shaped (depending on which side of the Atlantic Ocean you come from) shows an asteroid-like lightcurve, whereas dwarf planet (134340) Pluto (see Fig. 13.11) does not.

This figure shows the lightcurve of Pluto obtained by the author in the summer of 2009 and a surface map of the dwarf planet derived from Hubble Space Telescope images.

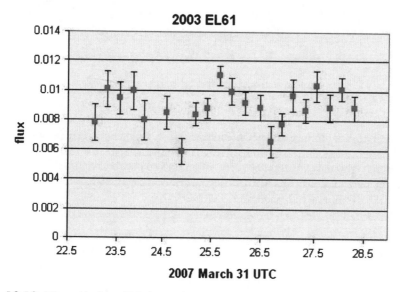

Fig. 13.10. Lightcurve of dwarf planet (136108) Haumea (Credit: John Saxton).

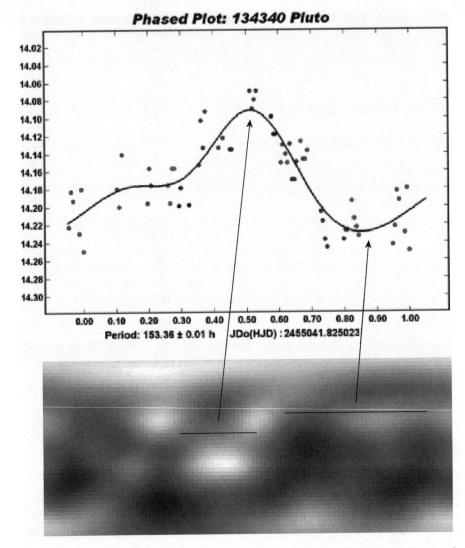

Fig. 13.11. Lightcurve aligned with surface map of dwarf planet (134340) Pluto (Credit: *MPO Canopus*/Alan Stern, Southwest Research Institute/Marc Buie, Lowell Observatory/NASA/ESA).

Closer to our home planet, Dr. Richard Miles used the 2.0 m (80 in.) robotic Faulkes Telescope South to image NEA 2008 HJ during April 2008. Figure 13.12 shows the lightcurve for this asteroid, which turned out to have the shortest known rotation period, 42.67 ± 0.04 s, of any natural Solar System body, the previous record, 78 s, being held by asteroid 2008 DO_8.

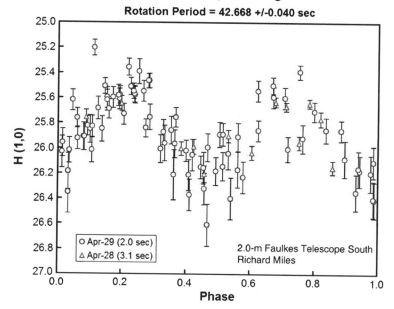

Fig. 13.12. Lightcurve of NEA 2008 HJ, the fastest known rotator in the Solar System (Credit: Richard Miles).

Conclusion

There is one repository for astrometry of asteroids and dwarf planets, and that is the Minor Planet Center. With photometry it is not so simple. Astrometric data sent to the Minor Planet Center *can* be accompanied by accurate photometry, but if your photometry is not so good then it is best not to include it in your report or the whole report may be rejected. Otherwise lightcurves can be submitted for publication in the *Minor Planet Bulletin*, uploaded to the CALL website via the Lightcurve Parameter Submissions Page, or sent to Raoul Behrend at Observatoire de Geneva (he should be contacted for the required format). National, regional, and local astronomical organizations may have their own files and/or databases and will usually welcome reports from their members. Where your observations are made as a result of a specific request, your data should of course be sent to the originator of that request.

By imaging an asteroid close to opposition you can obtain both a lightcurve and a phase curve and determine the object's absolute magnitude as well as its rotational period. Phase curves and absolute magnitude? Read the next chapter!

Chapter 14

Absolute Magnitude

When you look at an asteroid through the eyepiece of a telescope, or on a CCD image, it is a rather unexciting point of light. However by analyzing a number of images, information as to the nature of the asteroid can be gleaned. As we have seen in the previous chapter, frequent measurements of magnitude over periods of several minutes for fast rotators, or hours or even days for very slow rotators, can be used to generate a lightcurve. Analysis of such lightcurves yields the rotational period, shape, and pole orientation of the asteroid (Chap. 13). As described in Chaps. 10 and 11, measurements of position (astrometry) can be used to calculate the orbit of the asteroid and thus its distance from Earth and the Sun at the time of the observations. By combining the results from photometry and astrometry the absolute magnitude, H, and the slope parameter, G, can be derived. (It is quite common for G to be given a nominal value of 0.015.)

The H–G magnitude system, adopted by the International Astronomical Union in 1985, was developed for the purpose of predicting the magnitude of an asteroid as a function of solar phase angle. In addition the diameter of the asteroid can be calculated from H if its albedo is known or a value is assumed.

Theory

Both theory and practical work are covered in this chapter, so first we will explain some of the terminology used when exploring this particular aspect of asteroid observation and why 'absolute' may not be quite that. The remainder of the chapter will include practical advice, an example, and an update on professional activity in this area.

Apparent visual magnitude (V): the magnitude of an asteroid when observed and measured visually or with a CCD camera employing a suitable method to extract V (the response of the Johnson–Cousins V – visible light – filter approximates that of the human eye; therefore a visual magnitude, V, is equivalent to a Johnson–Cousins visible, V, magnitude).

Reduced magnitude, $H(\alpha)$: V with the influence of distance removed, e.g., relating solely to the phase angle α (and any variations in brightness due, for example, to the

R. Dymock, *Asteroids and Dwarf Planets and How to Observe Them*,
Astronomers' Observing Guides, DOI 10.1007/978-1-4419-6439-7_14,
© Springer Science+Business Media, LLC 2010

shape of the asteroid as covered in the previous two chapters). It assumes that the asteroid is 1 AU from both the Sun and Earth and is calculated using the equation

$$H(\alpha) = V - 5\log(r\Delta)$$

where (Fig. 14.1):

V = observed magnitude; r = distance of the asteroid from the Sun; Δ = distance of the asteroid from the Earth; α = the phase angle (Sun–asteroid–Earth angle).

Absolute magnitude, H: the V band magnitude of an asteroid if it was 1 AU from Earth and 1 AU from the Sun and fully illuminated, i.e., at zero phase angle (actually a geometrically impossible situation). H can be calculated from the equation:

$$H = H(\alpha) + 2.5\log[(1-G)\,\Phi_1(\alpha) + G\Phi_2(\alpha)]$$

where $\Phi_i(\alpha) = exp\left\{-A_i\left(tan^{1/2}\alpha\right)^{Bi}\right\}$ i=1 or 2; A_1=3.33, A_2=1.87; B_1=0.63 and B_2=1.22; α = the phase angle in degrees.

Thus at zero phase angle and with r = Δ = 1 AU, H = H(α). The various magnitudes mentioned above are average values, as the instantaneous value can vary by typically 0.5 magnitudes due to the rotation of the asteroid. The equation for calculating absolute magnitude is not valid for phase angles greater than 120° and is best used at much smaller values, e.g., 20° or less. This is not so much a problem when considering Main Belt asteroids, as their maximum phase angle does not usually exceed the latter value; however NEAs present more of a problem, as their maximum phase angles can approach 180°, which can lead to incorrect magnitude predictions. Both these alignments are shown in Fig. 14.2 – (21) Lutetia being a Main Belt asteroid and 2007 AG an Aten class NEA.

Slope parameter, G: relates to the opposition effect, which is a surge in brightness, typically 0.3 magnitudes, observed as the object approaches opposition. Its value depends on the way light is scattered by particles on the asteroid's surface. The roughness of the surface and the size, shape, and porosity of the particles all affect such

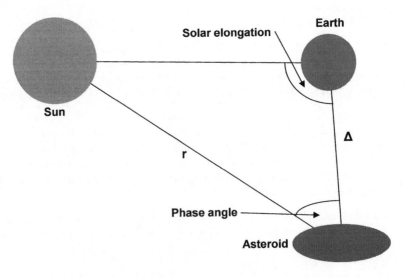

Fig. 14.1. Sun–Earth–asteroid relationship (Diagram by the author).

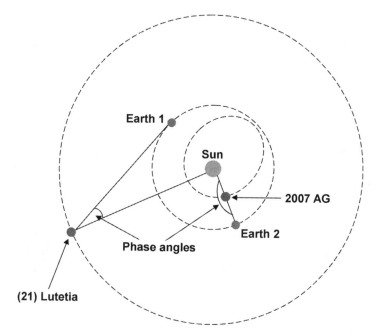

Fig. 14.2. Phase angles of Main Belt and near-Earth asteroids (Diagram by the author).

scattering. G is known for only a small number of asteroids; hence for most asteroids a value of 0.15 is assumed. Larger values of G result in larger (fainter) values of H.

Geometric albedo: the ratio between the brightness of a planetary body, as viewed from the Sun, and a white, diffusely reflecting sphere of the same size and at the same distance. Zero is used for a perfect absorber and one is applied to a perfect reflector – typical values for asteroids being between 0.05 and 0.25. Although G does vary with albedo the latter cannot be used to predict the former, as all asteroids with similar albedos do not have similar surfaces.

Phase curve: a graph of reduced magnitude vs. phase angle.

Phase coefficient, β: the slope of the linear portion of the phase curve between 10° and 20° of phase angle.

Not Quite Absolute

Unlike a star, the absolute magnitude, H, of an asteroid and the slope parameter, G, can have more than one value, and thus the quoted values are usually an average over several oppositions. The value of the absolute magnitude can be affected by the position of the object's rotational axis (aspect angle); for example we may see a typically egg-shaped asteroid end-on at one opposition and side-on at another (see Fig. 14.3). As a result the asteroid will appear to be brighter at Opposition 1 than at Opposition 2, and hence its value of absolute magnitude will be greater (less numerically, of course, which can be confusing).

The angle between an asteroid's spin axis and the ecliptic (obliquity) can also affect brightness at opposition and hence absolute magnitude, as shown in

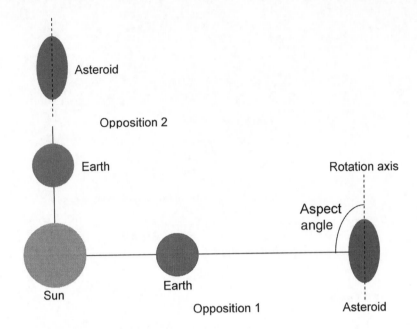

Fig. 14.3. Dependency of the absolute magnitude on aspect angle (Diagram by the author).

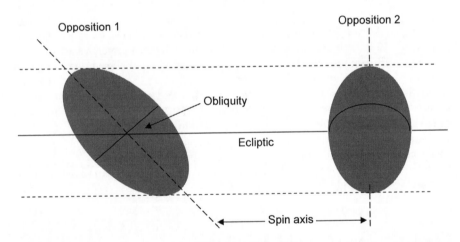

Fig. 14.4. Dependency of the absolute magnitude on obliquity (Diagram by the author).

Fig. 14.4. If we were to view the asteroid as it appears in the diagram at Opposition 1 we would see a larger profile than at Opposition 2, and hence the asteroid would appear brighter and its value of absolute magnitude would be greater. A fuller description of the pole position was given in the previous chapter.

Table 14.1 shows how H and G varied year by year for (1) Ceres and (8) Flora.

The magnitude relationships, and how reduced and absolute magnitudes are calculated, are summarized in Fig. 14.5.

The earlier definitions are shown graphically in Fig. 14.6, the apparent and reduced magnitude curves being constructed using data from the JPL Horizons website.

Table 14.1. Variation in values of H and G for (1) Ceres and (8) Flora

Asteroid	Year	H	G	Mean H/G
(1) Ceres	1990	3.29	0.08	
	1991	3.31	0.07	3.33/0.09
	1992	3.39	0.20	
(8) Flora	1990	6.42	0.27	
	1992	6.52	0.37	6.51/0.36
	1993	6.60	0.36	

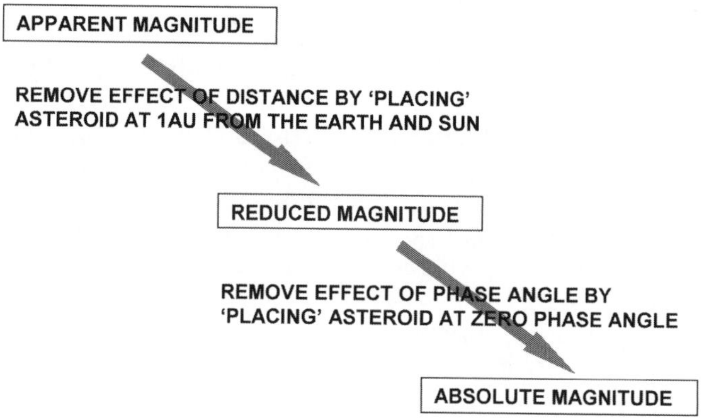

APPARENT MAGNITUDE
REMOVE EFFECT OF DISTANCE BY 'PLACING' ASTEROID AT 1AU FROM THE EARTH AND SUN
REDUCED MAGNITUDE
REMOVE EFFECT OF PHASE ANGLE BY 'PLACING' ASTEROID AT ZERO PHASE ANGLE
ABSOLUTE MAGNITUDE

Fig. 14.5. Magnitude relationships (Diagram by the author).

Reduced mag at 0° = Absolute mag = 7.35
Reduced magnitude
Opposition effect = 0.35 mag
Phase coefficient = 0.035 mag/deg
Apparent magnitude
Magnitude
Phase Angle
7
7.5
8
8.5
9
9.5
10
10.5
11
11.5
12
5
10
15
20

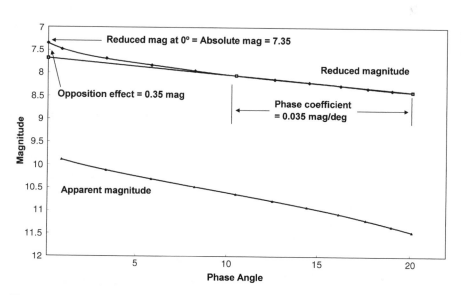

Fig. 14.6. Phase curve of asteroid (21) Lutetia (Diagram by the author).

Practical

What to Observe

To obtain a value for the absolute magnitude of an asteroid that object should be within 20° of its opposition point and, ideally, have a minimum phase angle of less than 1°. Suitable asteroid lists can be found in the following sources:

- On the Collaborative Asteroid Lightcurve Link (CALL website)
- In the *Minor Planet Bulletin*
- In the *Handbook of the British Astronomical Association*
- On the Magnitude Alert Project (MAP) website run by the Minor Planet Section of the Association of Lunar and Planetary Observers (ALPO)

The stated goal of the MAP is to obtain improved estimates of absolute magnitudes of asteroids. To achieve this goal it maintains a comprehensive list of asteroids with suspect values of absolute magnitude together with previous observational data and observation programs for the current year. Participants can receive e-mails containing details of recent observations and requests for further data.

The above-mentioned sources do list some asteroids that are bright enough for visual observation, but many will require the use of a CCD camera. Those observers actively participating in the Magnitude Alert Project use telescopes ranging from 200 to 730 mm (8–29 in.). Of course, what you can observe and derive a reasonably accurate magnitude estimate for is very much dependent on your location, affected as it may be by light pollution and poor seeing.

Not all asteroids make suitable targets, as some do not pass through 0° phase angle at opposition. For example the minimum phase angle of (23) Thalia was 7.8° at its 2007 opposition, due to the high inclination of its orbit (see Fig. 14.7). Don't give up on such asteroids, though, as their minimum phase angle will depend on the relationship between the position of their nodes with respect to Earth at opposition (orbital elements are covered in Chap. 2). If, for example, (23) Thalia, or any asteroid for that matter, were passing through its ascending or descending node at opposition then its phase angle at that time would be 0° or very close to that value.

Analysis

Analysis of observations is made easier by the availability of several software packages that will calculate values of H and G:

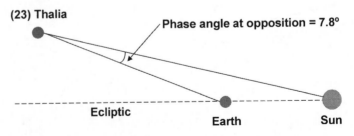

Fig. 14.7. Minimum phase angle of asteroid (23) Thalia (Diagram by the author).

- Comet/Asteroid Orbit Determination and Ephemeris Software (CODES) by Jim Baer, which will calculate orbital elements, H, G, and the asteroid's diameter (given a suitable value for its albedo) from astrometric and photometric observations.
- *Find_Orb*, from Project Pluto calculates orbital elements, an ephemeris, and H and G from observations in MPC format.
- *MPO Canopus* is used primarily for generating lightcurves, but it does include a utility for calculating H and G from photometric data providing that the orbital elements of the subject asteroid are known.

An Example: Asteroid (01909) Alekhin

As described in Chap. 12 this asteroid was imaged between March 24 and May 27, 2009, using the Sierra Stars Observatory Network robotic telescope. Figure 14.8 shows the phase curve calculated using the H/G Calculator in *MPO Canopus* – the scatter being due to variation in magnitude caused by the rotation of the asteroid as described in Chap. 12. When predicting magnitudes average values of observed magnitude are used, but for more detailed physical studies of an asteroid then H and G should be calculated using maximum values of observed magnitude.

Project Pluto's *Find_Orb* was used to calculate both orbital elements and H and G from the astrometric and photometric data obtained by analyzing the images using *Astrometrica* – the USNO-B 1.0 catalog for astrometry and the CMC14 catalog for photometry.

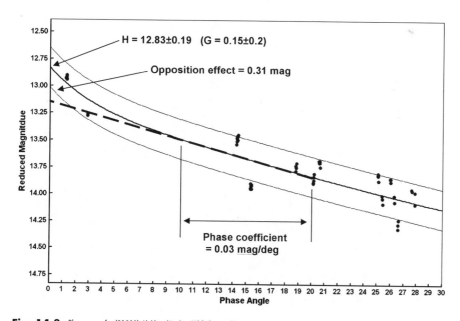

Fig. 14.8. Phase curve for (01909) Alekhin (Credit: *MPO Canopus*).

To run *Find_Orb*:

1. Load a file of observations in MPC format using the 'Open' button.
2. Double click on the file name that appears in the window below the 'Open' button.
3. The orbital elements that appear as a result of the above action can be further refined by clicking on the 'Auto-Solve' button one or more times. This should reduce the RMS error to less than 1 arcsec, which can be considered a reasonably good solution.
4. The results can be further refined by clicking on the 'Filter observations' button, which will exclude those observations with residuals greater than 1 arcsec.

Figure 14.9 shows the *Find_Orb* window after loading and calculating the orbital elements, H, and G for (01909) Alekhin. There are other facilities within this program that you can explore, but the simple explanation above is enough to get you started.

Table 14.2 compares the results from this exercise with the current data obtained from the Minor Planet Center (MPC). There is good agreement between both sets of figures, with the exception of the value of absolute magnitude. There are some discrepancies between MPC values for H and G and those obtained from recent observations. Providing data to allow the MPC values to be updated is, as mentioned above, one of the goals of the Magnitude Alert Project. Just to add to the confusion the MPC, AstDys, and JPL Horizons websites may offer different values for the absolute magnitude of an asteroid.

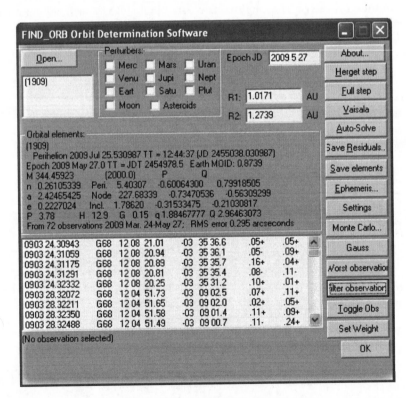

Fig. 14.9. Screen shot showing *Find_Orb* results for (01909) Alekhin (Credit: Project Pluto/Findorb).

Table 14.2. Comparison of *Find_Orb* and MPC results for the orbital elements of (01909) Alekhin

Name	Symbol	MPC notation		
Semi-major axis	a	a	2.4244561	2.42447621
Eccentricity	e	e	0.2226896	0.2226572
Inclination	i	Incl	1.78602	1.78615
Longitude of the ascending node	Ω	Node	227.68717	227.68416
Argument of perihelion	ω	Peri	5.41588	5.40462
Absolute magnitude	H	H	12.3	12.9
Slope parameter	G	G	0.15	0.15

If H is known it is possible to estimate the diameter of an asteroid by assuming a value for its albedo (see Chap. 4). Most Main Belt asteroids will have albedos between 0.05 and 0.25, in which case the diameter of (01909) Alekhin would lie between 5 and 7 km. Such estimates can be obtained, as this one was, using a conversion list on the MPC website or CODES, mentioned earlier in this chapter.

Publish!

As with all your observations they are of greater value if they are published. Particularly in the case of asteroids with suspect values of absolute magnitude the data should be sent (in the correct format, of course – Chap. 10) to the MPC and the MAP. Writing a short paper for the MPB and your local or national organization is another option and will always be welcome.

What of the Professionals?

This is not an area to which the professionals have previously directed a great deal of attention, but, in early 2009, a team of French and Italian astronomers devised a new method – using the European Southern Observatory's Very Large Telescope Interferometer (VLTI) – for measuring the sizes and shapes of asteroids that are too small or too far away for traditional techniques. Such traditional techniques include direct imaging using large ground-based telescopes with adaptive optics, space telescopes, simultaneous visible and infrared imaging, and radar measurements. The new method uses interferometry to resolve asteroids as small as about 15 km in diameter located in the main asteroid belt, 200 million km away. This technique will not only increase the number of objects that can be measured by a factor of several hundred, but more importantly, it will bring into reach small asteroids that are physically very different from the well studied larger ones.

The interferometric technique combined the light from two of ESO's 8.2 m (328 in.) telescopes, and the researchers applied this to imaging the Main Belt asteroid (234) Barbara. The observations revealed that this object has a peculiar shape – two bodies with diameters of 37 and 21 km forming either a peanut-shaped body or composed of two separate bodies in orbit round each other (see Fig. 14.10). If this asteroid is a binary then by combining the diameter measurements the

Fig. 14.10. Artist's impression of (234) Barbara (Credit: ESO/L. Calçada).

parameters of the orbits will allow astronomers to compute the mass and density of the two bodies. Having proven the validity of their new technique the team plan to start an observing campaign to study a large number of small asteroids.

Conclusion

Deriving absolute magnitudes has something to offer for visual observers, CCD imagers, and armchair observers who just like to experiment with available data. At one time ascertaining values of apparent magnitude from CCD images was not easy; however the new approach described in Chap. 12 (and an equivalent feature in the later versions of *MPO Canopus*) has made this a relatively simple task.

As previously mentioned, knowing the absolute magnitude, H, of an asteroid can lead to an estimate of its diameter. The size and shape of an asteroid can also be deduced from occultation data, as will be described in the next chapter.

Chapter 15

Occultations

Previous chapters have covered visual observing and various ways of imaging asteroids. As demonstrated in this chapter, observing occultations of stars by asteroids is a suitable activity for both methods and can yield results that tell us something of the nature of the asteroids involved. Very occasionally an occultation will occur that is visible to the naked eye, one example being the occultation of δ (delta) Ophiuchus by (472) Roma on July 8, 2010, which was visible from a number of European countries.

What Is an Occultation?

From time to time, during the course of its orbit around the Sun, an asteroid will pass between Earth and a star. As the asteroid moves along its orbit – A to B in Fig. 15.1 – the star will appear to suddenly dim or disappear altogether as seen along the track between C and D on Earth's surface. Such an event – an occultation – will typically last for just several seconds at any given location. The track width of the 'shadow' of the asteroid is narrower than that for a solar eclipse. Whereas the latter might typically be 200–300 km, for an asteroidal occultation the track width will be approximately equal to the diameter of the asteroid, e.g., 30–100 km. Usually there will only be a few potential observers along the track, so every observation, whether positive (occultation seen) or negative (occultation not seen), is important in defining the limits of the track and thus the size of the asteroid.

As has been mentioned previously some asteroids are binaries, and a significant proportion of stars are double or multiple. Where such objects are involved more than one disappearance/reappearance of the star may be observed. The majority of occultations feature Main Belt asteroids, but occasionally they do involve a trans-Neptunian object or the dwarf planet Pluto (the only such body known to possess an atmosphere). Such unusual events are worthy of a special effort.

What Can Occultations Tell Us?

Occultations present us with the only way of determining shape, apart from visits by spacecraft, radar imaging, or by large telescopes with adaptive optics. The shape of the asteroid is determined from the timings reported by observers, together with their locations. Figure 15.2 shows the results of observations, visual

R. Dymock, *Asteroids and Dwarf Planets and How to Observe Them*,
Astronomers' Observing Guides, DOI 10.1007/978-1-4419-6439-7_15,

Occultations

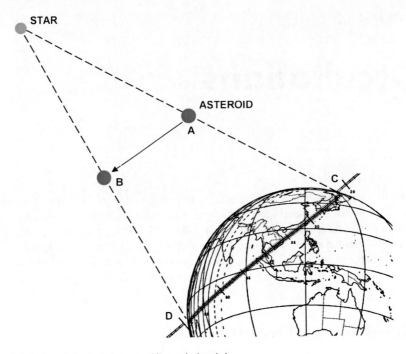

Fig. 15.1. An occultation of a star by an asteroid (Diagram by the author).

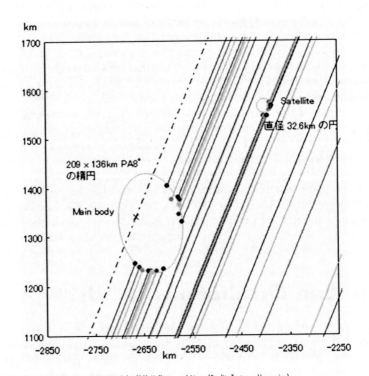

Fig. 15.2. Occultation of TYC 1886-01206-1 by (22) Kalliope and Linus (Credit: Tsutomu Hayamizu).

and using video cameras, by Japanese amateur astronomers of the occultation of star TYC 1886-01206-1 by asteroid (22) Kalliope and its satellite, Linus on November 7, 2006. The positions of the lines, or chords as they are usually referred to, relate to the locations of the observers and the breaks in the lines to the time and duration of the occultation. Some observers witnessed the occultation by the parent body and others by the satellite. The sizes of the two bodies deduced from the occultation data are shown in the figure and correspond well with other data. Figure 15.2 also illustrates the value of negative observations (those to the right of the main body and left and right of the satellite) in placing constraints on the size of both bodies.

In addition to yielding the size and shape of an asteroid, occultation data can:

– Help to refine the position of the star being occulted.
– Lead to a better defined orbit for the asteroid.
– Detect binary asteroids.
– Detect double stars, in particular close doubles.

As mentioned earlier, when the occulting body does not possess an atmosphere, an asteroid, or a moon, for example, then the light of the occulted star is instantaneously dimmed or cut off. When the former does possess an atmosphere, as is the case of dwarf planet (134340) Pluto, then the event happens more gradually due to refraction of the starlight through the atmosphere of the intervening object. This effect is shown in Fig. 15.3 – the fall and rise in magnitude both taking approximately 3 min.

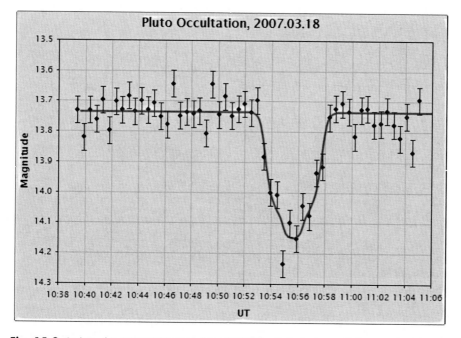

Occultations

Fig. 15.3. Occultation of star UCAC2 25823784 by dwarf planet (134340) Pluto (Credit: Chris Peterson, Cloudbait Observatory).

Occultation Predictions

Occultation predictions are available from several on-line sources, e.g., the International Occultation Timing Association (IOTA), Steve Preston's worldwide predictions website, the European Asteroidal Occultation Network (EAON), Eric Frappa's Euraster website, the Royal Astronomical Society of New Zealand, and astronomy magazines such as *Sky & Telescope*. Figure 15.4 shows part of a typical prediction from the EAON website, including the track, timing of occultation along the track, and known observing sites. The inner pair of parallel lines in Fig. 15.4 represents the predicted track, while the outer pair indicates the possible error in this prediction. In addition to that data the predictions often include (these vary from site to site):

– Finder charts at various scales
– Coordinates and magnitude of the star to be occulted
– Magnitude of asteroid
– Duration of and drop in magnitude of the star during the occultation
– Detailed track data, e.g., time of event at various locations (latitude, longitude) together with uncertainties in these data
– Angular distance of Sun and Moon from event and phase of the Moon

As mentioned earlier a prediction will include the expected drop in magnitude of the star to be occulted, so you can get an indication of what you can expect to see by searching for other stars in the FOV of similar brightness. For example if the target star is mag 9.2 and predicted to drop 4.4 mags, then look for stars at around mag 13.6. If this is fainter than your limiting telescopic magnitude the star will completely disappear. Although the predicted drop is usually several magnitudes it can, on

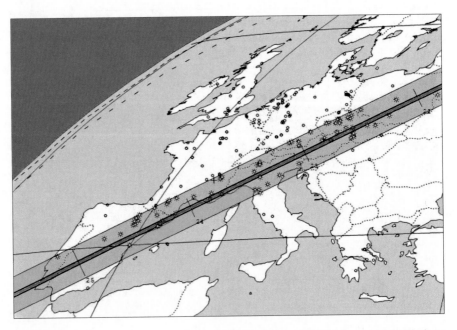

Fig. 15.4. Predicted ground track of the occultation of star 2UCAC 29564181 by asteroid (5337) Aoki on August 4, 2009 (Credit: Jean Schwaenen, EAON).

occasions, be less than this. It might therefore be worthwhile studying various fields of view to determine the minimum magnitude difference that you can easily detect.

Those who wish to generate their own predictions can download the free software package *Occult Watcher*. This Windows program searches various prediction sources for occultations that match the user criteria, such as star magnitude and event duration. Users can announce in advance which occultations they plan to observe, thus allowing worldwide coordination of events. You can of course generate your own finder charts using planetarium software such as *Megastar* or *Guide* and plotting the track of the asteroid for the period of the occultation. You may find that the ID of the star in the prediction is different from that in your planetarium software – for example star 2UCAC 29564181 is listed by *Megastar* and *Guide* as GSC 5205:358. However by consulting the finder chart included with the prediction and comparing the coordinates of the given UCAC star and the similarly positioned GSC star the correct star can be identified.

Predictions are by necessity only approximate, as the orbits of most asteroids and the positions of many stars are not known to the required level of precision to accurately define occultation tracks well ahead of time. The margin of uncertainty as far as the track center line is concerned is, thanks to the availability of *HIPPARCOS* astrometry, now probably better than ±50 km. The predicted times are usually accurate to within ±1 min. Astrometry is sometimes requested prior to the event, to allow the predicted track and time of occultation to be refined. Such last-minute changes are announced on the various websites mentioned earlier and can also be obtained by subscribing to e-mail alerts, e.g., *Sky & Telescope* magazine and mailing lists such as PLANOCCULT.

Observing an Occultation

If you are only a short distance from the predicted ground track and, of course, your equipment is easily portable, you might like to consider traveling to a suitable position to observe the event. Whereas many hundreds will travel vast distances to witness a solar eclipse, an asteroidal occultation doesn't excite quite so much interest, but some do make the effort, particularly in the United States and continental Europe. If you know of a local astronomical society with a permanent site on the track they may be willing to accommodate you. (This might encourage some of their members to observe the occultation – new adherents to this particular cause are always welcome.)

Visual Observing

If you are observing the occultation through a telescope or with binoculars then your preparations are as for visual observing, described in Chap. 8. Do give yourself enough time to set up your equipment and find the correct star field (an hour or so). Try and resist the temptation to drink anything between set-up and observation, for obvious reasons.

Spending several minutes staring at a star field through a telescope can be quite tiring, and the eyes start to play tricks. This is particularly the case if the target star is on the borderline of visibility due to its intrinsic magnitude, or partially obscured by a layer of thin cloud or a bright sky background; therefore practicing before the event, ideally a day or two before, is recommended. This will help you to determine whether or not you will have a clear view of the target star or whether some obstacle is blocking your view – particularly worth doing if the target star is low in the sky. It will also help you to identify the correct star field and, if the asteroid is bright enough, see it moving towards the target star. Such practice will also enable you to set the eyepiece in such a position that the event can be viewed from a comfortable sitting position if at all possible. This may be more of a problem for users of Newtonian reflectors than those with refractors or Schmidt–Cassegrains.

An alarm clock or timer with two alarms helps avoid unnecessary eyestrain. By setting the first alarm to 1 min before the predicted time and the second for the end of the suggested observing period you can increase your concentration nearer that time and know when to stop observing.

At the telescope, monitoring of the target star should commence some minutes before the predicted time and continue for at least the same period after it. (The occultation prediction data will usually include a suggested time period, and you can also check the timing for your location on the maps of the ground track.) At the instant of occultation the brightness of the star should be reduced by the predicted amount, which depends on the relative brightness of the asteroid and star being occulted. Beware also of secondary events, an occultation by a satellite of the minor planet, which may happen some time either side of the primary occultation. Although termed 'secondary,' such an event may also produce a dip in brightness virtually identical in magnitude to that of the primary event.

Two timings are necessary when a positive asteroidal occultation is observed – the disappearance or dimming of the star and the reappearance or brightening of the star being occulted. To accurately record these timings you will need:

- An accurate time-keeping source
- A timing device
- A measure of one's reaction time, or personal equation (PE)

Timing sources include radio controlled clocks, radio transmitters, GPS receivers, and Internet time servers. For the actual timing of the event a stopwatch or tape recorder together with an audible time signal broadcast by a radio station can be used. Ideally a stopwatch should have a lap timing facility so that you can record multiple events. To estimate your reaction time, or PE, cover the seconds, tenths, and hundredths of a second on a stop watch, then start it and stop it when the next full minute is indicated. The seconds display, which might typically be 0.4 s, is your PE. Repeating the exercise a number of times might give extremes of 0.7 and 0.3 s, with a most frequent value of 0.4 s, in which case you would quote your PE as 0.4 − 0.1/+0.3 s. Websites that allow you to calculate your PE are listed in Appendix B of this book. An example combining recorded times, time from a radio controlled clock, and PE is shown in Table 15.1.

Table 15.1. Recording and calculation of occultation timings

Activity	Stopwatch	Stopwatch time	Time of occultation	Time with PE applied
Star dims or winks out completely	Press Start	0	20:59:48 (21:06:00 − 6 min 12 s)	20:59:47.6 − 0.3/+0.1 (21:59:48 − 0.4)
Star returns to normal brightness	Press Lap	6.5 s	20:59:54.5 (20:59:48 + 6.5 s)	20:59:54.1 − 0.3/+0.1 (20:59:54.5 − 0.4)
Radio-controlled clock indicates a complete minute, e.g., 21:06:00	Press Stop	6 min 12 s		

Overall, taking all possible errors into consideration, an accuracy of ±0.3–0.4 s when making a visual observation gets a 'good' rating by the European Asteroidal Occultation Network.

Visual observations of occultations will enable you to gather useful data, but imaging gives you a permanent record of events and provides more accurate timings. An overview of such methods follows.

Video Recording

Occultations can be recorded using extremely sensitive, high-speed (25 frames/s) video cameras that can be fitted to a telescope in place of the eyepieces or used in conjunction with an image intensifier. Figure 15.5 shows UK amateur Andrew Elliott's set-up. The Watec 902A video camera has since been replaced by a more advanced Watec 120N hybrid version (with video rate imaging and integrating capability).

Used in conjunction with a GPS time inserter, the date and time being shown on the images, this set-up can record occultation timings to an accuracy of ±0.01 s. Frames from a video recording of the occultation of star TYC 4974-1069-1 by asteroid (121) Hermione are shown in Fig. 15.6, the faintest stars in these images being approximately 12th magnitude. The left-hand frame shows the star prior to the occultation and the right-hand one during the event when the star, occulted by the asteroid, has completely 'disappeared.'

Even a small telescope, when used in conjunction with an image intensifier, can enable objects as faint as those mentioned above to be imaged. U.S. amateur and IOTA secretary Richard Nugent uses a Collins I3 image intensifier attached to his

Fig. 15.5. Watec 902A hybrid video camera and image intensifier (Credit: Andrew Elliott).

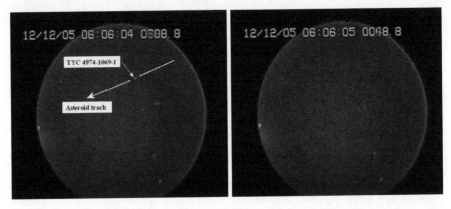

Fig. 15.6. Frames from a video of the occultation of star TYC 4974-1069-1 by asteroid (121) Hermione (Credit: Andrew Elliott).

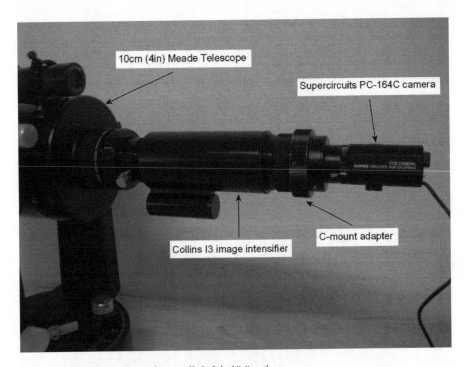

Fig. 15.7 Richard Nugent's video recording set-up (Credit: Richard N. Nugent).

10 cm (4 in.) Meade Schmidt–Cassegrain telescope (see Fig. 15.7). Used visually this set-up adds 1.5–2 magnitudes to his visual limit and, with a Supercircuits PC-164C camera, increases the limiting magnitude from 9.5 to 12.8.

Having captured the event on video the data can be extracted automatically using the free software package *Limovie* (**L**ight **M**easurement tool for **O**ccultation Observation using a **V**ideo recorder). *Limovie* will track the selected star plus up to two comparison stars in the images and record their pixel values and that for the sky background. The user has the ability to select the radius of the aperture

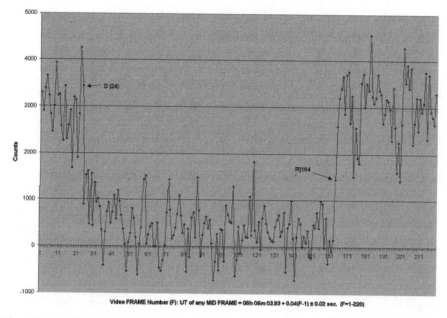

Fig. 15.8. Graph of the occultation of star TYC 4974-01069-1 by asteroid (121) Hermione on December 12, 2005 (Credit: Andrew Elliott).

over the selected stars and the annulus used to determine the sky background (as shown in Chap.12, Fig. 12.8). The data so obtained can be exported to a spreadsheet package and a graph of the event produced. Figure 15.8 shows the occultation of star TYC 4974-01069-1 by asteroid (121) Hermione on December 12, 2005. Each dot on the graph represents data from a single frame, and the disappearance and reappearance (reduction and increase in brightness) of the star can be clearly seen. The vertical axis shows pixel values and the horizontal axis video frame number, which can be converted into time.

The latest security CCD cameras have vastly improved sensitivity and can be plugged directly into a telescope without the need for an image intensifier. A 20 cm (8 in.) reflector with a typical video camera will allow objects as faint as magnitude 12 to be imaged.

CCD Imaging

CCD cameras take several seconds to download each image and therefore cannot accurately record an occultation if used in the 'normal' way as described in Chap.10. However if the telescope is operated in a fixed position, the star to be occulted will be recorded as a trail carrying a time versus position relationship for any change in magnitude during the exposure. Such drift scan images taken with a cooled CCD camera and properly calibrated provide a two-magnitude gain over low-light video cameras (with the exception of the integrating type). The positive or negative nature of the result is generally immediately apparent either in the image itself (a trace of the occultation of star UCAC 28565902 by asteroid (585) Bilkis on May 21, 2005, is shown in Fig. 15.9) or shortly after, when viewing a line profile (see Fig. 15.10).

Telescopes set up for occultation drift-scan observations require pre-pointing to a position adjacent to the target star so that the exposure can be taken while the star is within the field so that both ends of the trail are visible. A free software package, *ScanTracker*, can be used to coordinate these procedures.

Accuracy, being an all-important requirement of occultation timings, has been the stumbling block holding back drift-scan observations from mainstream activity. PC clocks alone are unlikely to provide sub-second accuracy. Beginning in 2003,

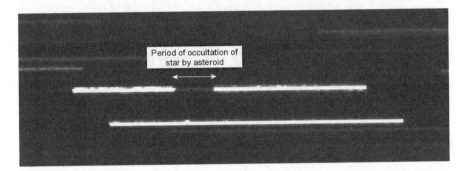

Fig. 15.9. Drift-scan image of a 10.8 magnitude star being occulted by 13th magnitude (585) Bilkis on May 21, 2005 (Credit: John Broughton).

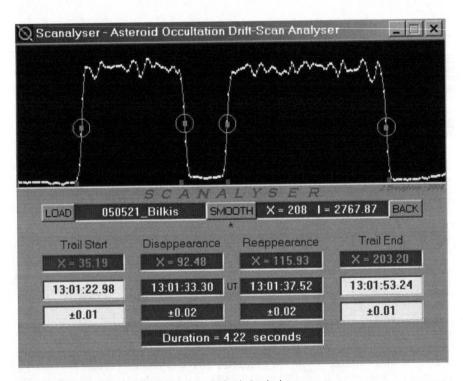

Fig. 15.10. Timing results of the (585) Bilkis occultation (Credit: John Broughton).

Australian minor planet observer John Broughton overcame this problem using audio recordings of either the shutter sound or RA drive-motor noise allowing points at the trail ends to be linked to UTC via a clock and short wave time signal. With exposure-timing inaccuracy taken out of the equation, timing precision then depends only on image quality, and accuracies of less than 0.1 s can be achieved.

For drift-scans showing a positive occultation, the analysis of the sound recordings can be done with audio software to derive the UTC time of the beginning and end of the exposure, which defines the end of the trail. Another free software package developed by John Broughton, *Scanalyzer*, displays the lightcurve, enabling disappearance, reappearance, duration, and timing uncertainty to be measured, taking into account optical distortion if necessary. Figure 15.10 shows the occultation by (585) Bilkis.

Electron Multiplying Charge Coupled Devices (EMCCDs) are far more sensitive but, at the present time, are rather expensive, even for the deepest amateur pocket.

Reporting

The various organizations mentioned in the Predictions section of this chapter require that reports are submitted using standard forms that are available from the websites of those organizations. If you were close to or on the predicted track then do report both positive and negative results, but if poor weather prevented observation then no report should be made. The example (see Table 15.2), with added explanatory notes, illustrates the data required to be reported and some of the difficulties faced.

Table 15.2. Example occultation report form

European Asteroidal Occultation Network (EAON)
International Occultation Timing Association/European Section (IOTA/ES)

Note: Please check the specific reporting requirements and required accuracy of each measurement for your area/organization, which may differ slightly from this example. Report forms preformatted for a specific event and region may be available on-line

1. Date	October 14, 2007	
	Star: TYC 0694-01184-1	Asteroid: (444) Gyptis
2. Observer	Name: David Storey	Phone:
	Address:	E-mail:
3. Observing station	Nearest city: Foxdale	Longitude: W 004 d 37 m 35.3 s
	Latitude: N 54 d 10 m 39.6 s	Altitude: 196 m
	Station: IAU 987 (The Isle of Man Observatory)	

Note: Latitude, longitude, and altitude can be obtained from, e.g., detailed maps or Google Earth, but the most accurate way is to use averaged (over 24 h at least) values from a GPS receiver. Some organizations do require you to report the source of these data. The 'IAU code' is the Minor Planet Center Observatory Code, described in Chap. 10. It is not necessary to obtain this for occultation work

4. Timing of events	Occultation recorded	
	(T)ype of event, (S)tart observation, (I)nterrupt-start, (D)isappearance, (B)link, (F)lash, (E)nd observation, (I)nterrupt-end, (R)eappearance, (O)ther (specify)	
	Personal Equation subtracted	P.E. − 0.6 on 'D' timing

(continued)

Table 15.2. (continued)

European Asteroidal Occultation Network (EAON)
International Occultation Timing Association/European Section (IOTA/ES)

Event Time (UTC)T: HH:MM:SS.ss	Comments
S: 00:48:20	Voice recording of time signal on MP3 recorder
O: 00:50:34	Star and asteroid merged into point source
D: 00:55:11.3	Combined asteroid/star faded. Star blinked twice, very quickly?
O: 00:55:23	Definitely faded
O: 00:55:43	Back to normal brightness?
O: 00:56:18	Hard to tell if the star/asteroid back to normal brightness
O: 00:56:33	Star/asteroid back to normal brightness, but did not see step in brightness
O: 00:59:31	Star and asteroid still merged into point source
O: 01:00:07	Possible asteroid and star separated

Note: No (R)eappearance was recorded as noted in section 8 below

O: 01:00:27	Star blinked out for a second or two? Possible poor seeing effect?
O: 01:01:54	Star and asteroid definitely separated
E: 01:02:05	End observation

Note: For a straightforward event (S)tart observation, (E)nd observation, and if the occultation is actually observed, (D)isappearance and (R)eappearance will suffice. Where the observation is less than straightforward added details, also mentioned in section 8 below, may help in the interpretation of the data

5. Telescope	Type: SCT Meade LX200 GPS	Aperture: 0.4 m
	Magnification: ×250 (40 mm Plossl + 2.5× Barlow)	Mount: Equatorial
	Motor drive: On	
6. Timing and recording	Timekeeping: Recorded time signal on MPS player. Stopwatch used on replay of MP3	Mode of recording: Visual observation recorded verbally onto MPS
7. Observing conditions	Atmospheric transparency: Good	Wind: Windy outside of dome, no effect inside observatory
	Star image stability: Poor	Minor Planet Visible: Visible until merged into star image
8. Additional comments	The actual occultation was well defined in the drop in magnitude of the combined star/asteroid image. This combined image did blink once or twice at this instant? Seeing effect? Very short duration blinks. I did not see a well defined increase of the star/asteroid image and would appear to have been a gradual rise? So I could not determine the duration of the occultation. Just after the asteroid was seen separated from the star, I did see the star blink out towards the end of the observation, but not sure if this was caused by poor seeing	

Results

As mentioned earlier in this chapter, the results from several observers can be combined to give an indication of the shape of the asteroid. In the case of (444) Gyptis, six observers obtained positive results and two were negative. As can be seen in Fig. 15.11 the negative results (chords 9 and 10) and the (D)isappearance (3) mentioned in the above report were useful in constraining the size of the asteroid. The dimensions of this asteroid were thus estimated to be 179.4 ± 3.9 km by 149.6 ± 2.7 km (112.1 ± 2.4 m by 93.5 ± 1.7 m).

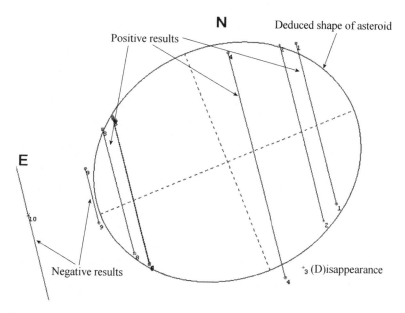

Fig. 15.11. Shape of asteroid (444) Gyptis deduced from occultation reports (Credit Eric Frappa, Euraster.net).

Conclusion

Although there is enough information in this chapter to enable you to become a reasonably competent occultation observer you may want to explore the subject further. For example, you might explore:

– *The Complete Guide to Observing Lunar, Grazing and Asteroid Occultations*, which can be downloaded for free from the IOTA website.
– The introduction, video observations, and Circulars sections on the EAON website.
– The Occultations section of the website of the Royal Astronomical Society of New Zealand.

Occultations are not just limited to the asteroid world, and you will probably have come across references to total and grazing lunar occultations. Observations of such events can be used to determine accurate positions of stars, separation, position angle, the relative magnitude of double stars, and, in the case of the latter, the profile of the lunar surface where the event occurred as seen from Earth. In addition skills developed observing asteroidal occultations can also be used to monitor eclipses, occultations, transits, and shadow transits of Jupiter's four Galilean satellites and the satellites of Saturn and Uranus. So you don't have to sit around waiting for the next asteroidal occultation at your location. You can try observing some of these other events. You will certainly not be alone if you involve yourself with this particular aspect of astronomy. There are around 300–400 regular occultation observers in Europe alone, many of whom use GPS time-stamped video systems.

However, while you are waiting the Internet beckons. It is possible to 'observe' asteroids, for example, by viewing images using NASA's Skymorph facility. The next chapter explains how to do this and examines past and future possibilities for the armchair observer.

Chapter 16

On-Line Image Analysis

You don't actually need any astronomical kit to observe and discover asteroids! All you require is Internet access to NASA's Skymorph facility. Before and after exploring that option we will make a brief detour into past and future opportunities.

If you do choose to go down this particular path and chance to make a discovery please make sure you follow the correct reporting procedures. If the process isn't fully explained on the project's website then do contact the source of the data or images for clarification.

The Past – Spacewatch Fast Moving Object (FMO) Project

This project ran for 2 years, from January 2004 to January 2006, during which time 43 asteroids were discovered by members of the public viewing images posted on-line by the University of Arizona's Spacewatch group. Fast-moving objects leave a short trail on images and, at the time, the automated detection software could not readily distinguish these from, for example, cosmic rays – hence the need for human intervention.

The review process involved the images being posted on the Internet, at which time, assuming you were logged on to the Spacewatch website, an alarm would sound on your PC. You then would download an image, check it for possible FMO tracks (they were unfortunately nearly always cosmic rays), and submit your findings to Spacewatch. Figure 16.1 shows the trail of asteroid 2006 AT$_3$ discovered by Hazel McGee on January 7, 2006. Discovery credit was given to both the on-line reviewer and the Spacewatch staff. Unfortunately this particular project is a thing of the past, but it was exciting while it lasted.

The Present – Skymorph

Today, using the SkyMorph website, you can access more than 650,000 images obtained by the Near Earth Asteroid Tracking (NEAT) program to search for asteroids (or any other moving or variable objects). A search for asteroids can be conducted in two ways: by time and position or by asteroid number.

R. Dymock, *Asteroids and Dwarf Planets and How to Observe Them*,
Astronomers' Observing Guides, DOI 10.1007/978-1-4419-6439-7_16,
© Springer Science+Business Media, LLC 2010

On-Line Image Analysis

Fig. 16.1. Trail of asteroid 2006 AT₃, discovered by Hazel McGee on a Spacewatch FMO image (Credit: Lunar and Planetary Laboratory, University of Arizona, Paul G. Allen Charitable Foundation, and NASA).

The method described here, developed by Marco Langbroek with assistance from Rob Matson and Stefan Kurti, can be used to discover asteroids and follow up on known or newly discovered objects. *Astrometrica* is used for astrometry and *Find_Orb* to calculate orbits. (The use of *Astrometrica* is covered in detail in Chap. 10 and Appendix C of this book.) Some knowledge of orbital elements is useful to verify that those output by *Find_Orb* are reasonable. (Chaps. 2 and 3 should help with this.)

Here is a short summary of the process:

1. Search for a moving object on a set of three images (referred to as a triplet).
2. Measure the object's positions and calculate a preliminary orbit.
3. Use the orbital elements to select further images.
4. Repeat steps 1–3 so that you have data from three nights.
5. Submit a report to the Minor Planet Center.

A detailed example follows using asteroid 33458 as the target, but first some notes on the Skymorph images. All images prior to December 2000 are from MPC station 566 (Haleakala); images after February 2001 are from station 608 (Haleakala), and all images with IDs ending in a, b, or c are from station 644 (Palomar). It is recommended that you use images from station 644, which are of a higher quality than those from stations 566 and 608, the latter containing many artifacts that make identifying moving objects extremely difficult.

1. Access Skymorph at http://skys.gsfc.nasa.gov/skymorph/skymorph.html and select 'Look for images by time and position.' Enter the coordinates: RA 07 27 04.8, Dec +19 11 19; time, 2002 01 01; and period, ±1 month; and then 'Submit query.' From the list of 'SkyMorph Matching Observations' produced, select a triplet – a group of three files obtained on the same day approximately 30 min apart and indicated by 'Y' in the right-hand column (Set 1 in Table 16.1).

Table 16.1. Skymorph images selected

Image set	Image ID	RA	Dec	Date and time
1	20011217091354a	07 27 11	+19 12 45	2001-12-17 09 14 24
	20011217093039a	07 27 10	+19 12 47	2001-12-17 09 31 09
	20011217094600a	07 27 10	+19 12 48	2001-12-17 09 46 30
2	011225095129	07 22 32	+20 02 30	2001-12-25 09 51 39
	011225100631	07 22 34	+20 02 31	2001-12-25 10 06 41
	011225102145	07 22 33	+20 02 50	2001-12-25 10 21 55
3	020114073421	06 58 27	+20 42 15	2002-01-14 07:34:31
	020114075023	06 58 26	+20 42 17	2002-01-14 07:50:33
	020114080533	06 58 26	+20 42 20	2002-01-14 08:05:43

The three images that make up the triplet will also have the same 'Observation Center' coordinates. In the 'Request Parameters' section set 'NEAT Pixels' to 1,100, select 'Show Singlets' and deselect any 'SkyView Comparisons.' Click on 'Request Images' and, when they have downloaded, save each image as a FITS file by right clicking on it and selecting 'Save Target As...' Make a note of the dates, times, RA, and Dec of the images, as you will need this information when processing the images with *Astrometrica*.

2. Rather than randomly selecting Skymorph images as a starting point you could select a known object from the various sources mentioned in Chap. 11 and then search for the associated images using the 'Look for an asteroid or other moving object' feature. Those images can then be searched for new objects using the *Astrometrica* blink option.

3. Open the downloaded files with *Astrometrica*, entering the correct image date and time as each image is opened. Make sure the configuration file used is that for the relevant MPC station. The station specific sections of these are shown in Table 16.2. (You may need to reset the directories to suit your circumstances.) To save the trouble of entering the data manually these configurations can be downloaded from Marco Langbroek's and Stefan Kurti's websites. Blink the three images and search by eye for any moving objects. Such movement should be approximately equal distance for each pair of images and consistent with the image sequence. Figure 16.2 is a stack of the three images that shows the movement of asteroid 33458 over a 32-min period.

4. On finding a moving object, you can check whether or not it is a known asteroid using Lowell Observatory's ASTPLOT feature. The reason for using ASTPLOT and not *Astrometrica*'s 'Known image overlay' option is that the Minor Planet Center's MPCORB file used by *Astrometrica* will be for the current epoch and not the epoch of these 'old' images. Select the correct station code from the list in ASTPLOT and fill in date, time, and image center coordinates of the middle image of the triplet. In addition change the limiting magnitude to '+22.5' and image vector to '1 h' and then click on 'Build Plot.' Figure 16.3 shows the resulting plot. (The output also includes a list of all the asteroids shown in the image.) Comparing this with the Skymorph images allows asteroid 33458 to be correctly identified.

5. If the asteroid detected on the images does not show on the plot then you may have discovered a new object. Whether new, known, or 'test' object, as in this case, measure its position on each of the three images using *Astrometrica*. To avoid losing data when starting *Astrometrica* rename the MPC report generated

Table 16.2. *Astrometrica* configurations for stations 566, 608, and 644

Parameter	Value		
Observing site			
MPC code	566	608	644
Longitude	156.2580 West	156.2580 West	116.8590 West
Latitude	20.7080 North	20.7080 North	33.3570 North
Height	3,000 m	3,000 m	1,000 m
CCD			
Focal length	2,186 mm ± 1.0%	2316.5 mm ± 5.0%	3075.7 mm ± 5.0%
Position angle	0.0 ± 2.0°	0.0 ± 10.0°	0.0 ± 10.0°
Pointing	±5.0′	±10.0′	±10.0′
Flip vertical	Checked	Checked	Checked
Auto-save FITS with WCS	Checked	Checked	Checked
Time in file header	Start of exposure	Middle of exposure	Middle of exposure
Offset	0.00 h from UT	0.00 h from UT	0.00 h from UT
Precision	1.00 s	1.00 s	1.00 s
Pixel width	15.0 μm	15.0 μm	15.0 μm
Pixel height	15.0 μm	15.0 μm	15.0 μm
Saturation	60,000	60,000	60,000
Color band	Red (R)	Red (R)	Red (R)
Exposure from FITS	Seconds	Seconds	Seconds
Program			
Aperture radius	3 Pixels	3 Pixels	3 Pixels
Detection limit	4.0 σ	4.0 σ	4.0 σ
Minimum FWHM	1.10 Pixels	1.10 Pixels	1.10 Pixels
PSF-Fit RMS	0.22	0.16	0.22
Search radius	0.36 Pixels	0.36 Pixels	0.36 Pixels
Background from	PSF	PSF	PSF
Plate constants	Linear fit	Cubic fit	Cubic fit
Astrometric limit	0.70Ð	0.70Ð	0.70Ð
Photometric limit	1.00 magnitude	1.00 magnitude	1.00 magnitude
Star catalog	USNO-B1.0	USNO-B1.0	USNO-B1.0
Upper limit	12.0 magnitude	12.0 magnitude	14.0 magnitude
Lower limit	19.0 magnitude	19.0 magnitude	20.0 magnitude
Reference matching			
Number of stars	50	120	100
Search radius	2.00 Pixels	2.00 Pixels	2.00 Pixels
Image alignment			
Number of stars	50	50	50
Alignment area	2,200 Pixels	300 Pixels	1,100 Pixels
MPC report	Include magnitude	Include magnitude	Include magnitude

and start a new one for each potential asteroid. If you believe the asteroid is a discovery then use your own provisional designation in the MPC report. Asteroid 33458 will be used to continue with this example.

6. Having found a moving object on the first set of images you now have to find the object on images from at least two other nights, one of which must be within a week of the first set. To do this use Project Pluto's *Find_Orb* software to calculate the object's orbital elements to enable you to predict where to look.

7. Start up *Find_Orb* and open the MPC report (or renamed file – in this case 33458. txt). Fill in a value for the perihelion distance that is typical of a Main Belt asteroid,

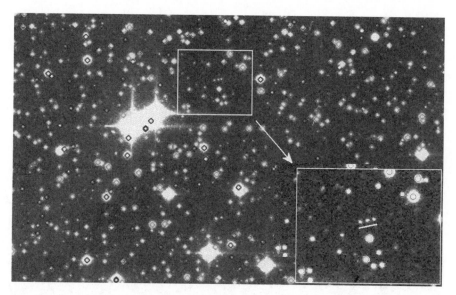

Fig. 16.2. Stack of three Skymorph images (image set 1) showing the movement of asteroid 33458 (Credit: Skymorph/*Astrometrica*).

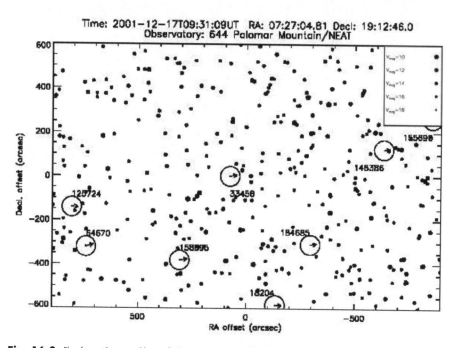

Fig. 16.3. Plot of asteroids generated by Lowell Observatory's ASTPLOT facility (Credit: Lowell Observatory Asteroid Services).

e.g., 2.7 AU. Click on 'Vaisala' (a Vaisala orbit is an orbit based on the assumption that the object was observed at perihelion). You will now get a set of estimated orbital elements (see Table 16.3, 1st pass) and an indication (RMS error, which should be less than 1.0) of the fit of the observations to this estimate. Main Belt asteroids are generally of low eccentricity and inclination, so check this is the case and that the semi-major axis (a) falls in the range of 2.1–3.3 AU.

Table 16.3. Orbital elements for asteroid 33458

Asteroid 33458	1st pass	2nd pass	3rd pass	MPC
Epoch (JD)	2452260.5	2452260.5	2452260.5	2452260.5
Eccentricity (e)	0.0790409	0.0559540	0.0895290	0.0859913
Perihelion distance (q)	2.30444390	2.17780502	2.11396578	2.1189946
Perihelion date (JD)	2451538.03	2452461.09	2452551.81	2452545.79
Longitude of ascending node (Node)	126.51817	129.27215	128.10311	128.20076
Argument of perihelion (Peri)	155.08662	32.36855	63.29714	61.13472
Inclination (Incl)	4.19738	3.19150	3.41810	3.39178
Absolute magnitude (H)	14.6	15.4	15.3	15.5

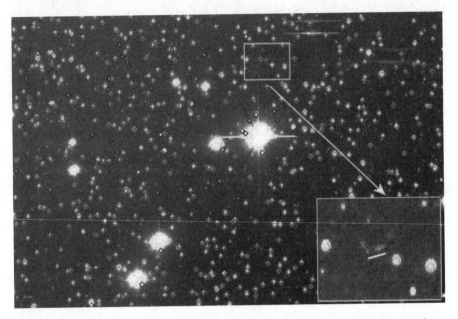

Fig. 16.4. Stack of three Skymorph images (image set 2) showing the movement of asteroid 33458 (Credit: Skymorph/*Astrometrica*).

8. Return to the Skymorph website and choose 'Look for an asteroid or other moving object.' Scroll down the page, enter the orbital elements from Table 16.3, 1st pass, and click on 'Submit Query.' Skymorph will return a list of images that may contain the object based on the preliminary orbit submitted. This list should include the original triplet of images; if it does not, you have made a mistake in the data input.

9. Select images from a second night as close to your first one as possible, download them, and use the *Astrometrica* blink feature to search for moving objects. If the preliminary orbit calculated by *Find_Orb* is a good estimate then you should be able to find the object, which will be of similar magnitude and moving in a similar manner to that from the first set. Checking with ASTPLOT confirmed that the moving object more or less in the center of the images (Fig. 16.4 shows part of the stacked image) was indeed asteroid 33458. Measure the position of asteroid 33458 on each of the images, add these new coordi-

nates to those measured on the first set of images, load them into *Find_Orb*, and calculate a new set of orbital elements (see Table 16.3, 2nd pass). To do this select all the 'Perturbers' and click on 'Autosolve' instead of 'Vaisala' as previously. Check that the residuals are less than 1.0 arcsec for each line of data. If they are not try using 'Herget step' or 'Full step.' If they are still much larger than the objects measured the two sets of images may not be the same, or there is a problem with your position measurements or associated dates and times.

10. Assuming all is well return to the Skymorph website and, as previously, choose 'Look for an asteroid or other moving object.' Scroll down the page, enter the orbital elements obtained by *Find_Orb* (see Table 16.3, 2nd pass), and click on 'Submit Query.' Skymorph will return a list of images that may contain the object. Check that the previous sets of images are listed. Download a set of three images from a third night dated as close as possible to the previous two sets (see Table 16.1, set 3). Checking with ASTPLOT the asteroid was found near the center of the images and its positions measured using *Astrometrica*. Figure 16.5 shows part of the stacked image. Using *Find_Orb* a further set of orbital elements were generated (see Table 16.3, 3rd pass). These could now be used to find further images and generate more positions if so desired. Table 16.3 also lists orbital elements for the same epoch obtained from the Minor Planet Center, and it can be seen that there is reasonable agreement between these and those generated from three sets of Skymorph images (3rd pass column).

11. Having assembled three nights of data, including two that are close together (within a week), you could, assuming they related to a discovery, submit them to the Minor Planet Center. (If your measurements relate to a recovery of a known object a single night's data can be reported.) Data from two nights can be submitted, but three are preferable and give more confidence that the

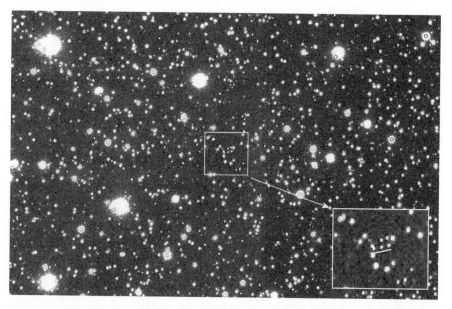

Fig. 16.5. Stack of three Skymorph images (image set 3) showing the movement of asteroid 33458 (Credit: Skymorph/*Astrometrica*).

observations are of the same object. *Astrometrica* outputs your data in the correct MPC format. Chapter 10 and Appendix D in this book are worth reviewing prior to submitting any reports to the MPC. As you will see in the example report below the OBServers are the NEAT staff, and you are the MEAsurer. As this was a known and numbered asteroid its correct designation, 33458, was used, but in the case of a discovery a preliminary designation of your own choosing, e.g., your initials and a number, should be included. Send the e-mail, as plain txt, not HTML, to mpc@cfa.harvard.edu. The body of the e-mail should not include anything other than the formatted report. Also note that observations from different stations should be submitted as separate items on your e-mail – see example below. The asterisk in the first line of data from station 644 indicates that this is the discovery image.

COD 644
CON *fill in your name,* *fill in your address* [*fill in your e-mail address*]
OBS R. Bambery, E. Helin, S. Pravdo, M. Hicks, K. Lawrence, R. Thicksten
MEA *fill in your name*
TEL 1.2-m Schmidt + CCD
ACK MPCReport file updated 2009.10.12 10:21:55
AC2 *fill in your e-mail address*
NET USNO-B1.0

33458	* C2001 12 17.38500 07 27 10.77 +19 12 44.1	18.5 R	644
33458	C2001 12 17.39663 07 27 10.19 +19 12 45.8	18.1 R	644
33458	C2001 12 17.40729 07 27 09.66 +19 12 47.4	18.2 R	644

----- end -----

COD 608
CON *fill in your name,* *fill in your address* [*fill in your e-mail address*]
OBS R. Bambery, E. Helin, S. Pravdo, M. Hicks, K. Lawrence, P. Kervin
MEA *fill in your name*
TEL 1.2-m reflector + CCD
ACK MPCReport file updated 2009.10.13 09:42:11
AC2 *fill in your e-mail address*
NET USNO-B1.0

33458	C2001 12 25.41087 07 20 01.49 +19 35 14.9	17.6 R	608
33458	C2001 12 25.42131 07 20 00.84 +19 35 17.2	17.5 R	608
33458	C2001 12 25.43189 07 20 00.20 +19 35 19.7	17.6 R	608
33458	C2002 01 14.31564 06 58 23.88 +20 42 11.0	17.8 R	608
33458	C2002 01 14.32677 06 58 23.13 +20 42 13.2	18.0 R	608
33458	C2002 01 14.33730 06 58 22.41 +20 42 15.4	17.8 R	608

----- end -----

The MPC will automatically acknowledge receipt of your e-mail and, assuming you have discovered a new object, will normally, within a day or so, send you an e-mail that includes your temporary designation and the MPC-assigned provisional designation. Then keep an eye on the Daily Orbit Updates where this object should be listed. Unfortunately you will no longer get official credit for asteroids discovered as described here, but you will have the satisfaction of knowing that you have made a discovery.

The Future - ?

The Large Synoptic Survey Telescope (LSST), on Cerro Pachón in Chilé, is due to become operational in 2015 and will search for potentially hazardous objects in the Solar System, i.e., Earth-approaching asteroids and comets, but many other deep-space objects will also be surveyed. Data, images, and alerts for specific types of events will be made available by the LSST team for analysis, and it is worth pointing out that objects brighter than magnitude 16 saturate the LSST's detectors, thus leaving an opportunity for amateur astronomers to discover the brighter asteroids. Other groups of professional astronomers may choose to involve a wider audience, as the increasing volume of data and images obtained becomes ever more difficult to manage.

A Final Word

What of the future for professional and amateur astronomers whose interest lies in asteroids and dwarf planets? Will the latter still be so named, and will the International Astronomical Union (IAU) come up with something that trips off the tongue a little easier than Small Solar System Bodies to describe asteroids and comets?

Can the *Messenger* spacecraft find an elusive Vulcanoid orbiting closer to the Sun than its primary target, Mercury? How will the *Dawn* mission to Ceres and Vesta change our ideas on the evolution of asteroids, dwarf planets, and the planets proper? What will the *New Horizons* spacecraft discover as it speeds past the dwarf planet Pluto and its attendant moons?

There is still much to understand about the parts of the Solar System to which this book relates. The distinction between asteroids and comets can seem blurred at times. Some NEOs, once thought to be asteroids, may actually be extinct comets, and icy comet-like bodies have recently been discovered in the asteroid Main Belt. Knowledge of the Edgeworth–Kuiper Belt is on a par with what was known about the Main Belt a century ago. It is quite probable that only a small percentage of its inhabitants – asteroids, dwarf, and not-so-dwarf planets – have been discovered and an even smaller number have had accurate orbits calculated. Further out, in the Oort Cloud, the majority of objects are believed to be comets, but a small, as yet unknown, proportion will probably turn out to be asteroids ejected from the region of the gas giant planets early in the life of the Solar System.

Even further afield, the discovery of extrasolar systems continues apace. Asteroid belts as well as planets have already been discovered around distant suns, and furthering our knowledge of these will help us understand how our own and other planetary systems formed. Progress in understanding the formation and evolution of planetary systems will need ever more complex computer models that can track the interactions of large numbers of bodies, the effects of collisions, and the slowly changing orbits of the planets.

To what use might we put asteroids? Mining these objects is one proposal, in particular those with a high metal content. The idea seems farfetched, but who knows? It has been suggested that suitable asteroids could be transported into orbits around Earth using similar methods to those that have been put forward to deflect possible impactors – so we had better get the mathematics right!

If the day ever comes when an asteroid 'with our name on it' is discovered will we be in a position to protect ourselves? Knowing the exact composition of such an Earth-approaching object – solid, rubble pile, or a dead comet – will be key to how we go about the task.

R. Dymock, *Asteroids and Dwarf Planets and How to Observe Them*,
Astronomers' Observing Guides, DOI 10.1007/978-1-4419-6439-7_17,
© Springer Science+Business Media, LLC 2010

The question is often asked whether amateur astronomers still have a role to play in this age of large, and soon even larger, telescopes and automated asteroid search programs, including the Panoramic Survey Telescope and Rapid Response System (Pan-STARRS) and the Large Synoptic Survey Telescope (LSST). Of course we will! As we hope this book has demonstrated, there are just too many asteroids and dwarf planets, known and unknown, and ways of observing them for amateurs to run out of things to do. Relationships between professional and amateur astronomers are in good health in, for example, the fields of astrometry, photometry, and occultations – long may they remain so. Many activities cross national borders, and this is certainly the case with asteroid astronomy. The ability to work together, for example to observe fast-moving near Earth asteroids can prevent such objects becoming lost as soon as they are discovered. Without the Internet the rapid dissemination of information would have been impossible, and asteroid astronomy would not be what it is today. Bad data as well as good can flow equally rapidly, so please do take the necessary steps to ensure your results are the latter before letting the world know.

Amateur capabilities are increasing all the time as telescopes become much more sophisticated, the range of imagers continues to grow and improve, and more powerful software and more accurate astrometric and photometric catalogs become available. Amateur astronomers have begun to probe the EKB, so further discoveries in that region may not be entirely in the hands of the professionals. Objects there are not only faint but move very slowly against the background stars. Following them requires a great deal of telescope time – something the professionals find it hard to obtain. It is to be hoped that professional search programs, as did the Spacewatch Fast Moving Object (FMO) Project, will make images available on-line for amateurs to peruse. Delicate electronic kits and the telescopes themselves do need protecting from the elements and are best left set up in their observing location, so an observatory is becoming a necessity. However this need not be large enough to accommodate the observer, as activities can be conducted over a network from a comfortable warm room. Observing at an even greater distance, using remote robotic telescopes operated over the Internet is becoming more common and less expensive. This offers tremendous opportunities for those who suffer from light pollution or do not have, or are unable to access, a suitable observing site.

As we learn new skills and make interesting observations there are a number of ways we can share them: forwarding measurements to the recognized worldwide organizations such as the Minor Planet Center (MPC) and the International Occultation Timing Association (IOTA); publishing methods and results in national and international journals; contributing to the many on-line interest groups and mailing lists; and making presentations at local, national, and international meetings.

Don't let all this talk of imaging, robotics, and software, though, put you off. There is tremendous satisfaction to be had observing visually using a relatively simple telescope, star-hopping to an asteroid, drawing the field of view, and being able to discern the asteroid's motion over an hour or so. It doesn't happen very often, but to see a bright NEO moving quite rapidly among the stars is mind-blowing.

As you would on any journey, if you are unsure of your way stop and ask. There are many out there who can, and will, be only too happy to help you. If you stick with it then you in turn will be able to repay those favors. You don't necessarily have to fund your hobby entirely from your own pocket, either. National and international organizations do support well thought out projects, e.g., the Planetary

Society's Shoemaker NEO grants. Whatever your interest – visual observation, imaging, using your own telescope or remote robotic facility, utilizing your own images or those stored in on-line repositories, searching on Google Earth for impact craters or, for the mathematically minded, calculating orbits using available software or even from first principles using Gauss's method (with which he determined the orbit of Ceres), we wish you success and hope this book will help you to pursue your particular interest.

Appendix A

Amateur and Professional Organizations

If a website addresses given here is no longer current, a search on the organization's name should lead to its new address.

Table A.1. Amateur and professional organisations

Name	Address
American Association of Variable Star Observers	http://www.aavso.org/
— Observing manual	http://www.aavso.org/observing/programs/ccd/manual/CCD_Manual_2010.pdf
Association of Lunar and Planetary Observers (ALPO)	http://alpo-astronomy.org/index.htm
— Minor Planets Section	http://www.alpo-astronomy.org/minor/
— Magnitude Alert Project	http://www.alpo-astronomy.org/minor/MAP_database_1.htm
— Minor Planet Bulletin	http://www.minorplanetobserver.com/mpb/default.htm
The Association of Space Explorers	http://www.space-explorers.org/
The Astronomer	http://www.theastronomer.org/index.html
Astronomical League	http://www.astroleague.org/index.html
— Asteroid Club	http://www.astroleague.org/al/obsclubs/asteroid/astrclub.html
— A Guide to Asteroid Observing	http://www.astroleague.org/al/obsclubs/asteroid/astrcobs.html
— What's up Doc ?	http://www.astronomyclub.org/wud.htm
B612 Foundation	http://www.b612foundation.org/
British Astronomical Association (BAA)	http://britastro.org/baa/
— Asteroids and Remote Planets Section	http://britastro.org/asteroids/
European Asteroidal Occultation Network (EAON)	http://astrosurf.com/eaon/
French Astronomical Society (Société Astronomique de France — SAF)	http://astrosurf.com/planetessaf/index_en.htm
— Occultations, Eclipses and Transits	http://astrosurf.com/planetessaf/occultations/index_en.htm
International Astronomical Union (IAU)	http://www.iau.org/
International Meteor Organisation (IMO)	http://www.imo.net/
International Occultation Timing Association (IOTA)	http://www.lunar-occultations.com/iota/iotandx.htm
Italian Organisation of Minor Planet Observers (Union Astrofili Italiani — UAI)	http://asteroidi.uai.it/
— Follow-Up Astrometric Program (FUAP)	http://asteroidi.uai.it/fuap.htm
The Meteoritical Society	http://www.meteoriticalsociety.org/
Orbit@home	http://orbit.psi.edu/
Planetary Society	http://www.planetary.org/home/
Royal Astronomical Association of New Zealand	http://www.rasnz.org.nz/

Appendix B

Resources

Listed here are books, websites, and mailing lists that will help the reader to further his or her knowledge of the topics covered in this book. Meteorites and comets are included in the 'Books' section, since the former, more often than not, originate from asteroids, and an interest in one form of small Solar System body often leads on to another.

We haven't attempted to list telescopes or imagers, as there are just too many combinations (and too many manufacturers and suppliers) to do so, as you will have seen from the examples in this book and the multiplicity of advertisements in astronomy magazines and on the Internet. To help you make your choice do talk to the more experienced amateurs, go to your local astronomy group, or contact your national organization, who will be able to give you advice.

Books

Asteroids and Dwarf Planets

Table B.1. Books — asteroids and dwarf planets

Title	Author/Editor	Publisher	Date
Asteroids III	W. F. Bottke Jr., A. Cellino, P. Paolicchi and R. P. Binzel	University of Arizona Press	2002
Asteroids, Comets and Meteorites: Cosmic Invaders of Earth	J. Erickson	Checkmark Books	2003
Asteroid Rendezvous: Near Shoemaker's Adventures at Eros	J. Bell and J. Mitton	Cambridge University Press	2002
Asteroids, Their Nature and Utilization	C.T. Kowal	Wiley	1996
Beyond Pluto	J. Davies	Cambridge University Press	2001
Dictionary of Minor Planet Names — 4th Edition	L. D. Schmadel	Springer	1999
Doomsday Asteroid — Can We Survive	D. W. Cox and J. H. Chestek	Prometheus	1996
Hazards due to Comets and Asteroids	T. Gehrels	University of Arizona Press	1995
The Hunt for Planet X — New Worlds and the Fate of Pluto	G. Schilling	Springer	2009
Introduction to Asteroids	C. J. Cunningham	Willman-Bell	1988
Mitigation of Hazardous Comets and Asteroids	M. Belton, T. H. Morgan, N. Samarasinha and D. K Yeomans	Cambridge University Press	1994
Planets Beyond: Discovering the Outer Solar System	M. Littman	Dover	2004
Pluto and Charon, Ice Worlds on the Ragged Edge of the Solar System	A. Stern and J. Mitton	John Wiley and Sons Ltd.	2005
Rogue Asteroids and Doomsday Comets	D. Steel	John Wiley and Sons Ltd.	1995
The Solar System beyond Neptune	M.A. Barucci, H.Boehnhardt, D.P. Cruikshank and A.Morbidelli	University of Arizona Press	2008
Target Earth	D. Steel	Time-Life	2000

Tools and Techniques

Table B.2. Books – tools and techniques

Title	Author/Editor	Publisher	Date
The Art and Science of CCD Astronomy	D. Ratledge	Springer	1997
The Backyard Astronomers Guide – 3rd Edition	T. Dickinson and A. Dyer	Firefly Books	2008
Digital Astrophotography – The State of the Art	D. Ratledge	Springer	2005
Electronic Imaging in Astronomy – Detectors and Instrumentation. 2nd Edition	I. S. McLean	Springer	2008
Fundamentals of Astrometry	J. Kovalevsky and K. Seidelmann	Cambridge University Press	2004
The Handbook of Astronomical Image Processing. 2nd Edition	R. Berry and J. Burnell	Willman-Bell	2005
Introduction to Astronomical Photometry. 2nd Edition	E. Budding and O. Dermircan	Cambridge University Press	2007
Introduction to Digital Astrophotography: Imaging the Universe with a Digital Camera	R. Reeves	Willmann-Bell	2005
Measuring Variable Stars Using a CCD Camera. A Beginner's Guide	D. Boyd	British Astronomical Association	2006
The New Amateur Astronomer	M. Mobberley	Springer	2004
A Practical Guide to CCD Astronomy	P. Martinez and A. Klotz	Cambridge University Press	1998
A Practical Guide to Lightcurve Photometry and Analysis	B. D. Warner	Bdw Publishing	2003
Setting up a Small Observatory: From Concept to Construction	David Arditti	Springer	2008
Stargazing Basics – Getting Started in Recreational Astronomy	P. E. Kinzer	Cambridge University Press	2008

Comets

Table B.3. Books – comets

Title	Author/Editor	Publisher	Date
Cometary science after Hale-Bopp, Vols. I & II	H. Boehnhardt, M. Combi, M. R. Kidger and R. Schulz	Kluwer Academic Publishers	2002
Comet of the Century – from Halley to Hale-Bopp	F. Schaff	Springer	1997
Cometography, A Catalogue of Comets Volumes 1, 2, 3 and 4	G. W. Kronk	Cambridge University Press	1999–2009
Comets and the Origin and Evolution of Life	P. J. Thomas, C. F. Chyba and C. P. McKay	Springer	2006
Comet Science	J. Crovisier and T. Encrenaz	Cambridge University Press	2000
David Levy's Guide to Observing and Discovering Comets	D. H. Levy	Cambridge University Press	2003
The Great Comet Crash: The Collision of Comet Shoemaker-Levy 9 and Jupiter	J. R. Spencer and J. Mitton	Cambridge University Press	1995
Great Comets	R. Burnham	Cambridge University Press	2000
Introduction to Comets	J. C. Brandt and R. D. Chapman	Cambridge University Press	2004
Observing Comets	N. James and G. North	Springer	2003

Meteor(s/ides/ites)

Table B.4. Books – meteor(s/ides/ites)

Title	Author/Editor	Publisher	Date
The Cambridge Encyclopaedia of Meteorites	O. R. Norton	Cambridge University Press	2002
David Levy's Guide to Observing Meteor Showers	D. H. Levy	Cambridge University Press	2008
Falling Stars: A Guide to Meteors and Meteorites	M. D. Reynolds	Stackpole Books	2001

Table B.4. (Continued)

Title	Author/Editor	Publisher	Date
Field Guide to Meteors and Meteorites	O. R. Norton and L. A Chitwood	Springer	2008
The Heavens on Fire	M Littmann	Cambridge University Press	1998
Meteorites — A Journey Through Space and Time	A. Bevan and J. de Laeter	Smithsonian Institute Press	2003
Meteorites and Their Parent Planets	H. Y. McSween	Cambridge University Press	1999
Meteors and How to Observe Them	R. Lunsford	Springer	2009
Meteorites, Messengers From Space	F. Heide and F. Wlotzka	Springer	1995
Meteorites — Their Impact on Science and History	B. Zanda and M. Rotaru	Cambridge University Press	2001
Meteors	N. Bone	Philip's	1993

Orbital Motion

Table B.5. Books — orbital motion

Title	Author/Editor	Publisher	Date
Fundamentals of Celestial Mechanics, 2nd Edition	J. M. A. Danby	Willman-Bell	1992
Orbital Motion	A. E. Roy	Institute of Physics	1988
Solar System Dynamics	C. D. Murray and S. F. Dermott	Cambridge University Press	1999
Theory of Orbit Determination	G. Gronchi and Andrea Milani	Cambridge University Press	2009

People

Table B.6. Books — people

Title	Author/Editor	Publisher	Date
Clyde Tombaugh, Discoverer of Planet Pluto	D. H. Levy	University of Arizona Press	1991
Shoemaker by Levy — The Man Who Made an Impact	D. H. Levy	Princeton	2000

Solar System

Table B.7. Books — solar system

Title	Author/Editor	Publisher	Date
The Compact NASA Atlas of the Solar System	R. Greeley and R. Batson	Cambridge University Press	1997
Encyclopedia of the Solar System — 2nd Edition	L.-A. McFadden, P. R. Weissman and T. V. Johnson	Academic Press	2007

Websites

An Internet search on 'asteroids' will yield over 650,000 hits – a few too many to include here!

Amateur Astronomers and Observatories

Table B.8. Websites – amateur astronomers and observatories

Name	Address
Ansbro E. – Kingsland Observatory	http://www.kingslandobservatory.com/Kingsland/Welcome.html
Birtwhistle P. – Great Shefford Observatory	http://www.birtwhistle.org/
Cahill A.	http://www.mountabbeydale.pwp.blueyonder.co.uk/
Črni Vrh Observatory	http://www.observatorij.org/
Fletcher J, – Mount Tuffley Observatory	http://www.jfmto.pwp.blueyonder.co.uk/
Higgins D. – Hunters Hill Observatory	http://www.david-higgins.com/Astronomy/index.htm
Hunter T. – Grasslands Observatory	http://www.3towers.com/
James N.	http://www.britastro.org/iandi/obsjames.htm
Kurti S.	http://www.skaw.sk/neattotalpage.htm
Langbroek M.	http://home.wanadoo.nl/marco.langbroek/
– Guide to recovering/discovering objects in the NEAT archive	http://home.wanadoo.nl/marco.langbroek/skymorph.html
Lowe A.	http://members.shaw.ca/andrewlowe/
McGaha J. – Sabino Canyon Observatory	http://www.3towers.com/sSabino/SabinoMain.asp
Mobberley M.	http://martinmobberley.co.uk/
Peterson C. – Cloudbait Observatory	http://www.cloudbait.com/
Observatorio Nazaret	http://astrosurf.com/nazaret/index.shtml
Robson M. – John J. McCarthy Observatory	http://www.mccarthyobservatory.org/
Saxton J.	http://www.lymmobservatory.net/ccd/ccd.htm
– LYMM Photometry Software	http://www.lymmobservatory.net/ccd/photometrysoftware/photsoft.htm
Shed of Science	http://home.earthlink.net/~shedofscience/index.html
Stevens J. and B. – Desert Moon Observatory	http://www.morning-twilight.com/dm448/
Sunflower Observatory	http://btboar.tripod.com/lightcurves/index.html
Sussenbach J	http://www.jsussenbach.nl/
Tucker R. A. – Goodricke-Piggott Observatory	http://gpobs.home.mindspring.com/gpobs.htm
Warner B. D. – Palmer Divide Observatory	http://www.minorplanetobserver.com/PDO/PDOHome.htm

Calendars, Catalogs, Finder Charts, and Surveys

Table B.9. Websites – calendars, catalogs, finder charts, and surveys

Name	Address
Aladin	http://aladin.u-strasbg.fr/aladin.gml
CalSky	http://www.boulder.swri.edu/ekonews/issues/past/n061/html/index.html
Heavens Above	http://www.heavens-above.com/
NASA Space Calendar	http://www2.jpl.nasa.gov/calendar/
Sloan Digital Sky Survey (SDSS)	http://www.sdss.org/segueindex.html
Small Main-Belt Asteroid Spectroscopic Survey (SMASS)	http://smass.mit.edu/smass.html
Space Telescope Science Institute (STScI)	http://www.stsci.edu/institute/
– Digitised Sky Survey (DSS)	http://archive.stsci.edu/cgi-bin/dss_form
United States Naval Observatory (USNO) Image and Catalog Archive Server	http://www.usno.navy.mil/USNO/astrometry/optical-IR-prod/icas
VizieR	http://vizier.u-strasbg.fr/viz-bin/VizieR
– Carlsberg Meridian Catalog 14 (CMC14)	http://vizier.u-strasbg.fr/viz-bin/VizieR?-source=I/304

Data and Information Sites

Asteroids and Dwarf planets in General

Table B.10. Websites – data and information sites, asteroids and dwarf planets in general

Name	Address
Atlas of mean motion resonances	http://www.fisica.edu.uy/~gallardo/atlas/
Asteroid (and Comet) Groups	http://sajri.astronomy.cz/asteroidgroups/groups.htm
Asteroid Comet Connection	http://www.hohmanntransfer.com/
Asteroid – Dynamic Site (AstDyS)	http://hamilton.dm.unipi.it/astdys/
Asteroid masses	http://home.earthlink.net/~jimbaer1/astmass.txt
BAA Asteroids and Remote Planets Section	http://britastro.org/asteroids/
The Belt of Venus	http://www.perezmedia.net/beltofvenus/
Description of the System of Asteroids	http://www.astrosurf.com/map/us/AstFamilies2004-05020.htm
Dwarf planets	http://www.gps.caltech.edu/~mbrown/dwarfplanets.html
How Gauss Determined The Orbit of Ceres	http://www.schillerinstitute.org/fid_97-01/982_orbit_ceres.pdf
CCD Observing Manual (American Association of Variable Star Observers – AAVSO)	http://www.aavso.org/observing/programs/ccd/manual/
IAU – definition of Pluto-like objects	http://www.iau.org/public_press/news/detail/iau0804/
IAU – planetary definitions	http://www.iau.org/public_press/news/detail/iau0603/
IAU – Pluto and the developing landscape of our Solar System	http://www.iau.org/public_press/themes/pluto/
Minor Planet Center (MPC)	http://www.minorplanetcenter.org/iau/mpc.html
— Guide to Minor Body Astrometry	http://www.minorplanetcenter.org/iau/info/Astrometry.html
— Minor Planet and Comet Ephemeris Service	http://www.minorplanetcenter.org/iau/MPEph/MPEph.html
— Minor Planet Electronic Circulars (MPECs)	http://www.minorplanetcenter.org/mpec/RecentMPECs.html
— MPChecker	http://scully.cfa.harvard.edu/~cgi/CheckMP
— NEO Confirmation Page (NEOCP)	http://www.minorplanetcenter.org/iau/NEO/ToConfirm.html
— Submission Information	http://www.minorplanetcenter.org/iau/info/TechInfo.html
Project Pluto	http://www.projectpluto.com/
— Minor Planet Groups	http://www.projectpluto.com/mp_group.htm
Skymorph	http://skyview.gsfc.nasa.gov/skymorph/skymorph.html
Small Body Distance Triangulation Calculator	http://heliospheric-labs.com/obs/calc.html
Updated Ephemeredes of Minor Planets	http://www.ipa.nw.ru/PAGE/DEPFUND/LSBSS/enguemp.htm
Uppsala Astronomical Observatory Planetary System Group	http://www.astro.uu.se/planet/
— Asteroids	http://www.astro.uu.se/planet/asteroid/

Binary Asteroids

Table B.11. Websites – data and information sites, binary asteroids

Name	Address
Asteroids with satellites	http://www.johnstonsarchive.net/astro/asteroidmoons.html
List des asteroids binaires	http://www.imcce.fr/page.php?nav=fr/ephemerides/donnees/binast
Minor Planet Satellite Database	http://www.astro.umd.edu/~dcr/badb/
Orbits of Binary Asteroids with Adaptive Optics	http://astro.berkeley.edu/~fmarchis/Science/Asteroids/

Impacts

Table B.12. Websites – data and information sites, impacts

Name	Address
Armagh Observatory	http://www.arm.ac.uk/home.html
— Near Earth Object Impact Hazard	http://star.arm.ac.uk/impact-hazard/
Earth Impact Database	http://www.unb.ca/passc/ImpactDatabase/
University of Arizona, Lunar and Planetary Laboratory	http://www.lpl.arizona.edu/
— Earth Impacts Effects Program	http://www.lpl.arizona.edu/impacteffects/
Holocene Impact Working Group (HIWG)	http://tsun.sscc.ru/hiwg/hiwg.htm
Impact Field Studies Group	http://web.eps.utk.edu/~faculty/ifsg.htm
Lunar and Planetary Institute	http://www.lpi.usra.edu/
— Terrestrial Impact Craters, Second Edition	http://www.lpi.usra.edu/publications/slidesets/craters/
NASA Ames Research Center, Asteroid and Comet Impact Hazards	http://impact.arc.nasa.gov/
— Torino Impact Scale	http://impact.arc.nasa.gov/torino.cfm
NASA JPL NEO Program	http://neo.jpl.nasa.gov/welcome.html
— Sentry Risk Table	http://neo.jpl.nasa.gov/risk/

Lightcurves

Table B.13. Websites – data and information sites, lightcurves

Name	Address
ALPO Minor Planets Section	http://www.alpo-astronomy.org/minor/
— Minor Planet Bulletin	http://www.minorplanetobserver.com/mpb/default.htm
Astronomical Society of Las Cruces	http://aslc-nm.org/AboutASLC.html
— Minor Planet Lightcurve Data	http://aslc-nm.org/Pilcher.html
Collaborative Asteroid Lightcurve Link (CALL)	http://www.minorplanetobserver.com/astlc/default.htm
Geneva Observatory — Asteroids and comets rotation curves	http://obswww.unige.ch/~behrend/page_cou.html
Koronis Family Asteroids Rotation Lightcurve Observing Program	http://www.koronisfamily.com/
Ondrejov Asteroid Photometry Project	http://www.asu.cas.cz/~ppravec/
Photometry of Asteroids at The Belgrade Astronomical Observatory	http://beoastrophot.awardspace.com/

Near-Earth Asteroids/Objects

Table B.14. Websites – data and information sites, near-Earth asteroids/objects

Name	Address
B612 Foundation	http://www.b612foundation.org/index.html
European Asteroid Research Node	http://earn.dlr.de/
Near Earth Objects Dynamics (NEODyS)	http://newton.dm.unipi.it/neodys/
Spaceguard	http://spaceguard.rm.iasf.cnr.it/SGF/INDEX.html
— Priority List	http://spaceguard.rm.iasf.cnr.it/SSystem/SSystem.html
Alain Maury	http://www.spaceobs.com/perso/recherche/whoiam.html
— NEO discovery statistics 2002	http://www.spaceobs.com/perso/recherche/NEA2002/NEA2002.htm
— NEO discovery statistics 2003	http://www.spaceobs.com/perso/recherche/NEA2003/NEA2003.htm
— NEO discovery statistics 2004	http://www.spaceobs.com/perso/recherche/NEA2004/NEA2004.htm
— NEO discovery statistics 2005	http://www.spaceobs.com/perso/recherche/NEA2005/NEA2005.htm
— NEO discovery statistics 2006	http://www.spaceobs.com/perso/recherche/NEA2006/NEA2006.htm
— NEO discovery statistics 2007	http://www.spaceobs.com/perso/recherche/NEA2007/NEA2007.htm

Occultations

Table B.15. Websites – data and information sites, occultations

Name	Address
Asteroidal Occultation Reports Database	http://sky-lab.net/cgi/occrep/submit/
Asteroid Occultation Updates (Steve Preston)	http://asteroidoccultation.com/
BREIT IDEAS Observatory	http://www.poyntsource.com/New/index.htm
Drift-scan timing of asteroid occultations	http://www.asteroidoccultation.com/observations/DriftScan/Index.htm
Euraster	http://www.euraster.net/
European Asteroid Occultation Network (EAON)	http://astrosurf.com/eaon/
International Occultation Timing Association (IOTA)	http://www.lunar-occultations.com/iota/iotandx.htm
RASNZ Occultation Section	http://occsec.wellington.net.nz/
SAF Occultations, Eclipses and Transits	http://astrosurf.com/planetessaf/occultations/index_en.htm

Phase Curves and Absolute Magnitude

Table B.16. Websites – data and information sites, phase curves and absolute magnitude

Name	Address
ALPO Minor Planets Section	http://www.alpo-astronomy.org/minor/
— Magnitude Alert Project (MAP)	http://www.astrosurf.com/map/index_us.htm
— Minor Planet Bulletin	http://www.minorplanetobserver.com/mpb/default.htm

Professional Astronomers and Observatories

Table B.17. Websites – professional astronomers and observatories

Name	Address
Arecibo Observatory	http://www.naic.edu/
Bland P.	http://www3.imperial.ac.uk/people/p.a.bland
Bottke W.	http://www.boulder.swri.edu/~bottke/Reprints/Reprints.html
Brown, M	http://www.gps.caltech.edu/~mbrown/
Bruton D.	http://www.physics.sfasu.edu/astro/asteroids/asteroids.html
Camarillo Observatory	http://www.camarilloobservatory.com/
European Southern Observatory (ESO)	http://www.eso.org/public/
Fitzsimmons A.	http://star.pst.qub.ac.uk/~af/About_Me.html
Jewitt D.	http://www2.ess.ucla.edu/~jewitt/David_Jewitt.html
Kaasalainen M.	http://www.rni.helsinki.fi/~mjk/asteroids.html
Kidger M.	http://www.observadores-cometas.com/
Klet Observatory	http://www.klet.org/
Kretlow M.	http://sky-lab.net/
Lowell Observatory	http://www.lowell.edu/
— Asteroid services	http://asteroid.lowell.edu/
Scotti J.	http://www.lpl.arizona.edu/~jscotti/
Skymapper	http://www.mso.anu.edu.au/skymapper/

Robotic Observatories

Table B.18. Websites – robotic observatories

Name	Address
Bradford Robotic Telescope	http://www.telescope.org/
Global Rent-A-Scope (GRAS)	http://www.global-rent-a-scope.com/
San Pedro de Atacama Celestial Explorations (SPACE)	http://www.spaceobs.com/index.html
Sierra Stars Observatory Network (SSON)	http://www.sierrastars.com/
Skylive	http://www.skylive.it/
SLOOH	http://www.slooh.com/

Search Projects

Table B.19. Websites – search projects

Name	Address
Catalina Sky Survey (CSS)	http://www.lpl.arizona.edu/css/
EURONEAR	http://euronear.imcce.fr/tiki-index.php?page=HomePage
Japanese Spaceguard Assocation	http://www.spaceguard.or.jp/ja/e_index.html
— Bisei Spaceguard Center	http://www.spaceguard.or.jp/bsgc_jsf/pamphlet/index.htm
— Kamisaibara Spaceguard Center	http://www.spaceguard.or.jp/bsgc_jsf/pamphlet/index.htm
Large Synoptic Survey Telescope (LSST)	http://www.lsst.org/lsst
Lincoln Near Earth Asteroid Research (LINEAR)	http://www.ll.mit.edu/mission/space/linear/
Lowell Observatory Near-Earth-Object Search (LONEOS)	http://asteroid.lowell.edu/asteroid/loneos/loneos.html
Near Earth Asteroid Tracking (NEAT)	http://neat.jpl.nasa.gov/
Panoramic Survey Telescope and Rapid Response System (Pan-STARRS)	http://pan-starrs.ifa.hawaii.edu/public/
Spacewatch	http://spacewatch.lpl.arizona.edu/
— Spacewatch Fast Moving Object Project	http://fmo.lpl.arizona.edu/FMO_home/index.cfm
UAO-DLR Asteroid Survey (UDAS)	http://solarsystem.dlr.de/SGF/earn/udas/

Software

Please note that some packages may contain features other than those indicated by the table headings. Electronic cameras (CCD, DSLR, and video) will come with their own software, but most, if not all, will work with other packages. For example you can use *Megastar* to point your telescope and *Astroart* to obtain the images.

Astrometry and Photometry

Table B.20. Websites – software, astrometry and photometry

Name	Address
AIP4WIN	http://www.willbell.com/aip/index.htm
Astroart	http://www.msb-astroart.com/
Astrometrica	http://www.astrometrica.at/
Iris	http://www.astrosurf.com/buil/us/iris/iris.htm
Mira AL	http://www.mirametrics.com/mira_al.htm
PinPoint	http://pinpoint.dc3.com/

Celestial Mechanics

Table B.21. Websites – software, celestial mechanics

Name	Address
Comet/Asteroid Orbit Determination and Ephemeris Software (CODES)	http://home.earthlink.net/~jimbaer1/
Find_Orb	http://www.projectpluto.com/find_orb.htm
OrbFit	http://adams.dm.unipi.it/~orbmaint/orbfit/
Solex	http://main.chemistry.unina.it/~alvitagl/solex/

DSLR and Video

Table B.22. Websites – software, DSLR and video

Name	Address
Astrostack	http://www.astrostack.com/
DeepSkyStacker	http://deepskystacker.free.fr/english/index.html
Registax	http://www.astronomie.be/registax/
Virtual Dub	http://www.virtualdub.org/

Integrated Packages

Table B.23. Websites – software, integrated packages

Name	Address
Diffraction Limited	http://www.cyanogen.com/index.php
— Cloud Sensor	http://www.cyanogen.com/cloud_main.php
— MaxDome II	http://www.cyanogen.com/dome_main.php
— Maxim DL	http://www.cyanogen.com/maxim_main.php
— MaxPoint	http://www.cyanogen.com/point_main.php
Minor Planet Observer	http://www.minorplanetobserver.com/
— Asteroid Viewing Guide	http://www.minorplanetobserver.com/MPOSoftware/MPOViewingGuide.htm
— Canopus	http://www.minorplanetobserver.com/MPOSoftware/MPOCanopus.htm
— Connections	http://www.minorplanetobserver.com/MPOSoftware/MPOConnections.htm
— LCInvert	http://www.minorplanetobserver.com/MPOSoftware/MPOLCInvert.htm

Observatory Control

Table B.24. Websites – software, observatory control

Name	Address
ACP Observatory Control	http://acp.dc3.com/index2.html
Omega Lab Astronomy Programs and Observatory Control	http://www.omegalab-atc.com/
Remote Administrator (RADMIN)	http://www.radmin.com/

Occultations

Table B.25. Websites – software, occultations

Name	Address
Light Measurement tool for Occultation Observation (LIMOVIE)	http://www005.upp.so-net.ne.jp/k_miyash/occ02/limovie_en.html
Occult Watcher	http://www.hristopavlov.net/OccultWatcher/OccultWatcher.html
Scanalyser, Scantracker	http://www.asteroidoccultation.com/observations/DriftScan/Index.htm

Planners and Planetarium Programs

Table B.26. Websites — software, planners and planetarium programs

Name	Address
Asteroid	http://amhugo.tripod.com/asteroid.html
Astroplanner	http://www.ilangainc.com/astroplanner/index.html
Guide	http://www.projectpluto.com/
Megastar	http://www.willbell.com/software/megastar/index.htm
Sat_ID	http://www.projectpluto.com/sat_id.htm
Skymap	http://www.marinesoft.co.uk/skymap

Time Synchronization

Table B.27. Websites — software, time synchronization

Name	Address
Chronos Atomic Clock Synchronizer	http://download.cnet.com/Chronos-Atomic-Clock-Synchronizer/3000-2350_4-10861687.html?tag=mncol%3blst
CNS Systems TAC32	http://www.cnssys.com/cnsclock/Tac32Software.php
Dimension 4	http://www.thinkman.com/dimension4/index.htm
TimeSync	http://www.ravib.com/timesync/

Space Agencies

Table B.28. Websites — space agencies

Name	Address
European Space Agency (ESA)	http://www.esa.int/esaCP/index.html
Japanese Aerospace Exploration Agency (JAXA)	http://www.isas.jaxa.jp/e/index.shtml
National AeroSpace Agency (NASA)	http://www.nasa.gov/
— Astrophysics Data System	http://adsabs.harvard.edu/index.html
— Jet Propulsion Laboratory (JPL)	http://www.jpl.nasa.gov/
— Asteroid Radar Research	http://echo.jpl.nasa.gov/
— Horizons	http://ssd.jpl.nasa.gov/horizons.cgi
— Near Earth Object Program	http://neo.jpl.nasa.gov/welcome.html
— What's Observable	http://ssd.jpl.nasa.gov/sbwobs.cgi
— Lunar and Planetary Science	http://nssdc.gsfc.nasa.gov/planetary/
— NEO Survey and Deflection. Analysis of Alternatives	http://www.nasa.gov/pdf/171331main_NEO_report_march07.pdf
— NASA Science	http://nasascience.nasa.gov/
— Planetary Data System (PDS)	http://pds.jpl.nasa.gov/
— Small Bodies Node	http://sbn.pds.nasa.gov/

Space Telescopes and Missions

Table B.29. Websites — space telescopes and missions

Name	Address
Dawn	http://dawn.jpl.nasa.gov/
Don Quijote	http://www.esa.int/esaCP/SEML9B8X9DE_index_0.html
Gaia	http://www.esa.int/esaSC/120377_index_0_m.html
Hayabusa	http://www.isas.jaxa.jp/e/enterp/missions/hayabusa/index.shtml

Table B.29. (Continued)

Name	Address
HIgh Precision PARallax Collecting Satellite (HIPPARCOS),	http://sci.esa.int/science-e/www/area/index.cfm?fareaid=20
Hubble Space Telescope	http://hubblesite.org/
MErcury Surface Space ENvironment GEochemistry and Ranging (MESSENGER)	http://messenger.jhuapl.edu/
Near Earth Asteroid Rendezvous (NEAR)	http://near.jhuapl.edu/
Near Earth Object Surveillance SATellite (NEOSSAT)	http://www.neossat.ca/
New Horizons	http://pluto.jhuapl.edu/
Origins Spectral Interpretation Resource Identification and Security (OSIRIS)	http://sse.jpl.nasa.gov/missions/profile.cfm?Sort=Alpha&Letter=O&Alias=OSIRIS
Rosetta	http://sci.esa.int/science-e/www/area/index.cfm?fareaid=13
SOlar and Heliospheric Observatory (SOHO)	http://sohowww.nascom.nasa.gov/home.html
Wide-field Infrared Survey Explorer (WISE)	http://wise.ssl.berkeley.edu/index.html

Mailing Lists

Table B.30. Mailing Lists

Name	Address
Minor Planet Mailing List	http://tech.groups.yahoo.com/group/mpml/
Planoccult	To subscribe contact Pierre Vingerhoets at Pierre.Vingerhoets@telenet.be or Jan Van Gestel at jan@vangestel.be

Appendix C

Papers

Included in this appendix are:

A paper entitled 'A method for determining the V magnitude of asteroids from CCD images' first published in the June 2009 issue of the *Journal of the British Astronomical Association.*

A Method for Determining the V Magnitude of Asteroids from CCD Images

Roger Dymock and Richard Miles

A contribution of the Asteroids and Remote Planets Section (Director: Richard Miles)

We describe a method of determining the V magnitude of an asteroid using differential photometry, with the magnitudes of comparison stars derived from *Carlsberg Meridian Catalogue 14* (CMC14) data. The availability of a large number of suitable CMC14 stars enables a reasonably accurate magnitude (±0.05 mag) to be found without having to resort to more complicated absolute or all-sky photometry. An improvement in accuracy to ±0.03 mag is possible if an ensemble of several CMC14 stars is used. This method is expected to be less accurate for stars located within ±10 degrees of the galactic equator owing to excessive interstellar reddening and stellar crowding.

The Problem

Differential photometry is fairly straightforward in that all the required data can be obtained from the images of the target object. Changing atmospheric conditions and variations in dimming due to altitude are likely to affect all stars equally and can thus be ignored. However, deriving magnitudes in this way is problematic in that comparison stars with known accurate magnitudes are few and far between, and it is unusual to find many, if any, such stars on a typical CCD image. So to obtain an accurate measure of the magnitude of the target asteroid, one must ordinarily resort to 'absolute' or 'all-sky' photometry. This approach is however much more complicated than differential photometry: the sky must be adequately clear (sometimes referred to as 'photometric'); standard stars need to be imaged (usually some distance from the target asteroid); at least one filter must be used; and extinction values, nightly zeropoints, airmass corrections and the like need to be taken into account.

Proposed Solution

Overview

From a single image or, preferably, a stack of several images, the Johnson V magnitude (centred on ~545 nm) of an asteroid may be obtained, accurate to about ±0.05 mag. All that is required is access to the *Carlsberg Meridian Catalogue 14* (CMC14) from which data the magnitudes of the comparison stars can be calculated. We set out here a method for deriving such V magnitudes of comparison stars in the range 10 < V < 15. It is possible that a more automated approach could be devised by someone with sufficient computer skills (see Postscript for significant developments in this respect since this paper was first written).

The CMC14 Catalogue

The Carlsberg Meridian Telescope (formerly the Carlsberg Automatic Meridian Circle) is dedicated to carrying out high-precision optical astrometry.[1] It underwent a major upgrade in 1999 March with the installation of a 2k × 2k pixel CCD camera together with a Sloan Digital Sky Survey (SDSS) r′ filter, the system being operated in drift-scan mode.[2] With the new system, the r′ magnitude limit is 17 and the positional accuracy is in the range 35–100 milliarcsec. The resulting survey is aimed to provide an astrometric and photometric catalogue in the declination range −30 to +50°. The CMC14 catalogue is the result of all the observations made between 1999 March and 2005 October in this declination band, except for a gap between 5 h 30 m and 10 h 30 m for declinations south of −15°. It contains 95,858,475 stars in the Sloan r′ magnitude range 9–17. The Sloan r′ band is in the red part of the spectrum centered at ~623 nm and having a bandwidth at half-maximum of 137 nm.

Importantly, a VizieR query[3] of the CMC14 catalogue returns both r′ magnitude data as well as the Two-Micron All-Sky Survey (2MASS) data comprising the J-band (1.25 μm), H-band (1.65 μm), and K_s-band (2.17 μm) magnitudes.[4] To be able to convert the known r′ magnitude of a star into a standard V magnitude, we must know the colour of the star. The difference in the J and K magnitudes provides such a measure of star colour.

Theory

In 2006, John Greaves analysed data for 696 stars (9.9 < V < 14.8) from the LONEOS photometric database produced by Brian Skiff.[5] He merged these with r′ magnitude data from the earlier Carlsberg CMC12 catalogue and with 2MASS data (0.00 < (J–K) < 1.00) and found that the relationship

$$V = 0.641 \times (J - K) + r' \qquad \text{(C.1)}$$

predicted the V magnitude with fair accuracy (standard deviation of 0.038 mag) when comparing V(Loneos) − V(calculated) using the above linear relationship.[6]

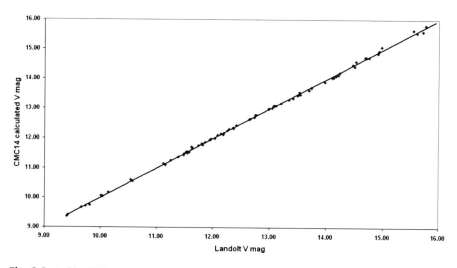

Fig. C.1. Landolt vs CMC derived V magnitudes

It was found that a limit has to be placed on the allowable values of (J–K) since the above relationship begins to break down for very red stars. From further studies, John Greaves and Richard Miles concluded that the choice of comparison stars should be limited to those with a value of (J–K) between 0.3 and 0.7. However, where there is only one comparison star in the image the allowable J–K range might have to be relaxed to between 0.2 and 0.8.

In a separate exercise, the V magnitudes of one hundred Landolt stars, the accepted standard for visible photometric calibration, were calculated using (C.1) above and a graph of those derived values plotted against the Landolt magnitudes.[7] Subsequent analysis of the data by the present authors led to a 'best-fit' modification of the above formula to

$$V = 0.6278 \times (J - K) + 0.9947 \times r' \qquad (C.2)$$

A graph of V magnitude derived from the CMC14 catalogue versus Landolt magnitude is shown in Fig. C.1. The mean error of the CMC14-derived V magnitudes was calculated to be 0.043 for stars brighter than V = 14.

Plotting residuals as shown in Fig. C.2 clearly illustrates the departure of the calculated magnitude from the Landolt values based on (C.2) above. Throughout the magnitude range, 9 < r' < 16, the average correlation shows no systematic trend away from linearity, only a slightly increased scatter at V > 15. Also shown plotted in Fig. C.2 are the differences between CMC-derived V magnitudes and those calculated from *Tycho-2* data using *GUIDE 8.0* software respectively.[8,9] The *Tycho* catalogue created as part of the *Hipparcos* space mission comprises about 2.5 million stars and has been used by many as a source of magnitude data. However, it can be seen from Fig. C.2 that there is significant scatter for the *Tycho* data at magnitudes fainter than about 10.5, showing that this commonly used catalogue is unsuitable for use as a source of accurate photometry for stars fainter than this limit. Unlike *Tycho* the CMC14 catalogue provides a good reference source for field calibration down to 15th magnitude.

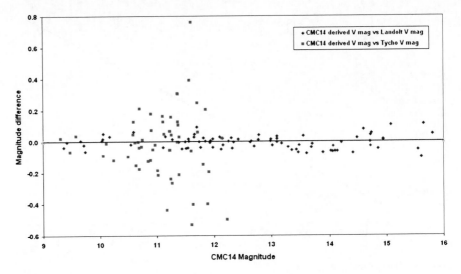

Fig. C.2. Residual plot of Landolt vs CMC14-derived magnitudes and Tycho V magnitudes

Practical Results

The equipment used by the authors is listed in Table C.1.

Table C.1. Instrumentation used

Item	Roger Dymock	Richard Miles
Telescope	Orion Optics 25-cm f/6.4 Newtonian on a German equatorial mount	Celestron 28 cm f/10 Schmidt–Cassegrain and Takahashi 6 cm f/5.9 refractor, both on the same German equatorial mount
CCD camera	Starlight Xpress MX516	Starlight Xpress SXV-H9
Filter	Johnson V	Johnson V, Cousins I and unfiltered

Example 1: Determining the V Magnitude of Asteroid (1467) Mashona

Imaging, calibration, and magnitude calculation described in this example were carried out by RD as detailed below. Asteroid *(1467) Mashona* imaged on 2007 October 17 is used as the example.

Guidelines

Here are some guidelines to follow if photometry of asteroids is planned:
- Choose asteroids higher than 25–30° altitude.
- Do not attempt to image objects that are too faint to ensure a sufficiently high signal-to-noise ratio (SNR), ideally >20: in RD's case using a 0.25 m telescope this means working on objects brighter than V = 15. The SNR can be improved by stacking multiple images.

- Choose an exposure time which avoids excessive trailing (say by more than 2 or 3 pixels) due to the motion of the asteroid.
- Check the stacked images to ensure that the asteroid and comparison stars do not become saturated. Where necessary stack in 'Average' mode to ensure saturation is avoided.
- Check for suitable comparison stars in the field of view (FOV): in RD's case at least two stars brighter than V = 14. A planetarium program such as *MegaStar*[10] or *Guide*[9] is useful for doing this and to avoid inadvertently using known variable stars as comparison stars.
- Access the CMC14 catalogue to verify that the chosen comparison stars are of a satisfactory colour, i.e. have a (J–K) value between 0.30 and 0.70, or 0.20–0.80 if there is only one comparison star in the FOV.

If these guidelines are not followed much time can be wasted 'chasing' unsuitable asteroids and determining magnitudes which may have considerable uncertainty.

Imaging

Twenty images of an asteroid were obtained using an exposure time of 30 s. To avoid trailing when stacking images, the maximum time interval from start of the first image to the end of last image should not exceed a value defined using a formula proposed by Stephen Laurie, i.e. the maximum interval in minutes equals the full width at half maximum (FWHM) of a star image in arcseconds divided by the rate of motion of the asteroid in arcseconds per minute.[11]

For *(1467) Mashona* the motion amounted to 0.039 arcsec/m as obtained from the Minor Planet Center's Ephemeris Service,[12] and the FWHM was 4 arcsec giving a total duration of: 4/0.039 = 102 m therefore trailing would not pose a problem using a sequence of five images obtained over a 5-m interval.

Calibration frames consisting of five dark frames, five flat-fields, and five flat-darks (dark frames having the same exposure as the flat-fields) were taken for each imaging session.

Image Processing

Master dark frames and master flat-fields were generated using the software, *Astronomical Image Processing for Windows (AIP4WIN)*.[13] Calibration frames were median-combined and saved. Images containing the asteroid were calibrated and stacked, typically 3–5 images in each stack, to improve SNR, using the *Astrometrica* software.[14] The stacked image was measured using *AIP4WIN*. A whole series of images can be processed in one batch run using the 'Multiple Image Photometry Tool' facility. The first image in a series is used to set up the analysis which can then be automatically applied to all images in that series.

The task now is to utilise data from the CMC14 catalogue for each of the comparison stars to derive its V magnitude and, together with the *AIP4WIN* data, calculate the V magnitude of the asteroid.

Accessing the CMC14 Catalogue Using Aladin

A convenient tool for carrying out the analysis is the online software, *Aladin Sky Atlas*.[15] Using this facility it is possible to display and align one's own image, a Digital Sky Survey (DSS) image and the CMC14 catalogue, in chart form as shown in Fig. C.3.[16]

Performing an astrometric calibration will align the CMC14 chart with both images. Clicking on the relevant star displays the CMC14 data for that star as shown near the bottom of Fig. C.3. The data for each star can then be copied to, for example, an Excel spreadsheet.

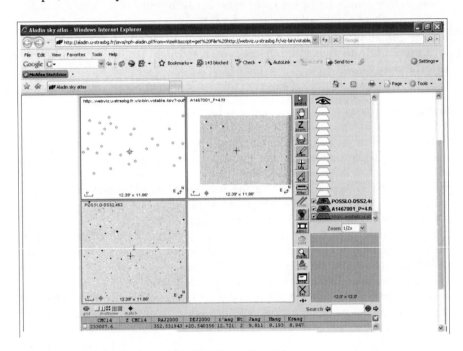

Fig. C.3. CMC14 chart (*top left*), RD's image (*top right*) and DSS image (*bottom left*)

Magnitude Calculation

The magnitude of the asteroid was calculated with the aid of a spreadsheet using the relevant CMC14 catalogue values and (C.2) as set out below. The results are shown in Table C.2.

- The values r', J and K for each star obtained from CMC14 are given in columns F–H
- (J–K) is calculated in column I
- The V magnitude, in column J, is calculated from the formula
- $V = 0.6278 \times (J - K) + 0.9947 \times r'$
- The instrumental magnitudes, v, (from *AIP4WIN*) for each of the comparison stars, C1–C4, and for the asteroid are given in column B
- (v–V) for each comparison star is calculated in column C, rows 3–6
- The mean of (v–V) for all four comparison stars is calculated in cell C7
- The V magnitude of the asteroid is calculated in cell D9 by subtracting the mean (v–V) from the instrumental magnitude, v, of the asteroid

Table C.2. Calculation of V magnitude for asteroid *(1467) Mashona*

Col.:	A	B	C	D	E	F	G	H	I	J	K	L
Row												
1	AIP4WIN data				CMC14 data							
2		v	(v–V)	Calc	GSC No.	r'	J	K	J–K	V	Error	Used
3	C1	−6.47	−19.67		1726-1112	12.72	9.81	8.95	0.86	13.20	0.02	No[a]
4	C2	−7.34	−19.80		1726-1088	12.31	11.45	11.11	0.34	12.46	0.01	Yes
5	C3	−6.99	−19.76		1726-1073	12.48	11.10	10.54	0.56	12.77	0.01	Yes
6	C4	−6.41	−19.77		1726-1377	13.16	11.94	11.50	0.44	13.36	0.02	Yes
7	Mean		−19.78									
8	Ast	−6.57									0.02	
9	Asteroid V mag			13.21								
10	Error – Imaging ±			0.03								
11	Error – Cat ±			0.03								
12	Error – Total ±			0.04								

[a] Comparison star C1 was not used in the calculation because it was found to possess a J–K colour index outside the acceptable 0.30–0.70 range described earlier

Errors

The errors are calculated as described below.

Catalogues

Calculating the uncertainty in the results is an important part of the exercise as we are seeking to demonstrate that this methodology is significantly better than other available options (*Tycho-2*, *UCAC-2*, *USNO-A2.0*, etc.).[17,18] Many variable star observers have used the *Tycho* catalogue as a source of comparison star data. The All-Sky Survey (TASS), for example, is calibrated against *Tycho-2* stars.[19] *Guide* states that 'In most cases the precision provided by *Tycho* is much greater than all earlier catalogues. About the only case in which *Tycho* data would be ignored is if *Hipparcos* data is available instead'.[20] Although *Hipparcos* is indeed a very accurate catalogue for V photometry, it contains only 118,209 stars, most of which are brighter than V = 9 and so are effectively too bright and too sparsely distributed to be of use as photometric reference stars in most cases.

One problem with using *Tycho* data even for the brighter stars is that there are not likely to be many such stars on any particular image. For example, a 12 × 8 arcmin image typically contains one Tycho star. A larger FOV will contain more but at high galactic latitudes, where stellar densities are much lower than the average, it is common not to have any *Tycho* stars in the FOV of a CCD image. By comparison, CMC14 contains almost *40 times* as many stars as *Tycho*: hence its suitability as a potential source of comparison stars. Table C.3 lists the uncertainties quoted for the magnitude of individual stars in CMC14 depending on their brightness.

Table C.3. CMC14 photometric uncertainty

r'	Δr' (mag)
<13	0.025
14	0.035
15	0.070
16	0.170

It can be seen that fainter than r′ = 15, the error in r′ becomes significant. It is therefore best if we only use r′ mags between 9 and say 14.5. In an area relatively devoid of comparison stars (say if only one is present), you might have to resort to stars fainter than r′ = 14.5. For example, the average of a group or ensemble of say four or five r′ = 15.5 stars may be as good as a single star brighter than r′ = 14.5. Typically, the magnitude error for an ensemble of reference stars is given by the catalogue error divided by (square root of no. of stars used *minus* 1). Since (C.2) also includes the (J–K) term, errors in the 2MASS catalogue will also contribute towards the uncertainty in the derived V magnitude. Note that for greater accuracy, the colour range of comparison stars is restricted to those with 0.3 < (J–K) < 0.7.

Signal-to-Noise Ratio

The errors (standard deviation or sigma) as reported by *AIP4WIN* represent only some of the sources of uncertainty in that they are based on the photon counts for the star and sky background and do not take into account many other factors affecting accuracy such as variations in sky transparency, reference catalogue errors, accuracy of flat-fields, etc. Using the method described in the AAVSO *CCD Observing Manual Section 4.6*, [21] the instrumental magnitudes of stars over a range of magnitudes were measured from 30 images using *AIP4WIN*. The standard deviations or sigmas of the instrumental V magnitudes were compared with the *AIP4WIN*-derived sigmas, and an equation for the relationship derived, namely:

$$Actual\,error = 1.13 \times AIP4WIN\ error + 0.007 \qquad (C.3)$$

It should be noted that (C.3) will be different for different combinations of telescopes and CCD cameras and observers might wish to perform their own calculations.

For the example here, the total SNR contribution to the error for the three reference stars and the asteroid (see Table C.2) is given by:

$$\sqrt{0.01^2 + 0.01^2 + 0.02^2 + 0.02^2}$$

Combined Error

Assuming the component errors are independent, the overall uncertainty (1-sigma standard deviation) is the square root of the sum of the squares of the total SNR error and the reference catalogue error, viz.:

$$Combined\,error = \sqrt{0.03^2 + 0.03^2} = \pm 0.04\,mag$$

Colour Transformation

If we wish to combine measurements with those of other observers, it is often necessary to measure the small differences between your own telescope/CCD/filter system and the standard filter passbands in order to correct or transform measurements to a standard magnitude system (in this case Johnson V). The transformation coefficient for RD's set-up was ascertained with the help of the BAA's Variable Star Section's publication, *Measuring Variable Stars using a CCD camera – a Beginner's Guide.*[22] This method, described in Appendix 6 of that document, uses *Hipparcos* red-blue pairs which are close enough to appear on the same CCD image.

The colour correction is that value which would have to be added to the calculated V magnitude to convert it to a standard V magnitude. Corrections were calculated using the (B–V) values shown in parentheses typical of C-type asteroids (0.7), S-type (0.9), and blue (0.2), red (1.0) and G-type stars (0.6). We ignore extremely red and blue stars in this exercise and where possible use an ensemble of stars to derive the magnitude of the asteroid so that colour corrections tend to balance out. For RD's system, the average correction that would have to be applied proved to be less than 0.01 mag and can therefore be disregarded.

Example 2: Determining the V Magnitude of Asteroid 2000 BD$_{19}$

CMC14-Derived V Magnitudes Compared with Absolute Photometry Based on Hipparcos Reference Stars

Imaging, calibrations and magnitude calculations described in this example were carried out by RM. As a real-life example, the images have been selected from a campaign to observe the unusual asteroid *2000 BD$_{19}$* between the dates, 2006 Jan 24–Feb 10. This object is unusual in that it has the lowest perihelion distance (0.092 AU) of any object in the Solar System for which orbits are accurately known, whilst its aphelion distance is virtually the same as that of Mars.

The observing methodology involved imaging the asteroid in a large telescope (FOV = 8.5 × 11 arcmin) with no filter and imaging the same comparison stars together with several stars from the *Hipparcos* Catalogue using two wide-field telescopes (FOV = 63 × 86 arcmin), one equipped with a V filter, the other a Cousins Ic filter. A type of absolute photometry was carried out on the wide-field images, each frame being calibrated in terms of the mean zeropoint, (v–V) of the *Hipparcos* stars adjusted to zero V–Ic colour index using previously derived transformation coefficients for each filter passband. Knowing the V–Ic colour of the comparison stars, it was possible to calculate the V magnitude of these stars to an accuracy of about ±0.015 mag. It is then possible to compare these directly-measured values with CMC14-derived V magnitudes.

Imaging

For this example, images made on two nights were used. A series of wide-field images were made through filters, namely a stack of 20 × 90 s V- and Ic-filter exposures on 2006 Jan 25 00:01–00:33UT, and a stack of about 80 × 60 s V- and Ic-filter exposures on 2006 Jan 29/30 23:42–01:31UT. In the former case, the field was centered near RA 11 h 35 m, Dec. +35.5° and four *Hipparcos* stars were used as standards (HIP 56516, 56568, 56671 and 56799). In the latter case, the field was near RA 11 h 18.7 m, Dec. +41° and four other *Hipparcos* stars were used as standards

(HIP 55101, 55182, 55194 and 55503). A number of suitable comparison stars were identified and their V magnitude determined, values of which are listed in the second column of Tables C.4 and C.5. The corresponding CMC14 data were obtained from the catalogue and Equation (C.2) was employed to derive the V magnitude of each of the stars (14 in all).

Table C.4. Comparison between measured and calculated V magnitude using Equation (C.2). Comparison stars in the field of asteroid *2000 BD19* on 2006 January 24/25

Star ref.	V Mag	r' Mag	J Mag	K Mag	J–K Colour	V_{calc}	$V - V_{calc}$
a	13.78	13.46	12.39	11.96	0.43	13.74	0.04
b	13.28	13.06	12.16	11.83	0.33	13.27	0.01
c	13.62	13.21	11.32	10.56	0.76	13.70	−0.08
d	13.65	13.37	12.00	11.46	0.54	13.72	−0.07
e	13.88	13.66	12.62	12.23	0.39	13.91	−0.03
f	13.19	12.82	11.34	10.70	0.64	13.23	−0.04
g	13.05	12.70	11.18	10.54	0.64	13.11	−0.06
						Mean	−0.031
						St. dev	0.044

Table C.5. Comparison between measured and calculated V magnitude using Equation (C.2). Comparison stars in the field of asteroid *2000 BD19* on 2006 January 29/30

Star ref.	V Mag	r' Mag	J Mag	K Mag	J–K Colour	V_{calc}	$V - V_{calc}$
d	10.86	10.66	9.80	9.49	0.31	10.86	0.00
g	12.67	12.45	11.68	11.41	0.27	12.62	0.05
i	10.82	10.60	9.60	9.24	0.36	10.83	−0.01
j	11.85	11.58	10.60	10.25	0.35	11.80	0.05
s	12.70	12.45	11.41	11.03	0.38	12.69	0.01
t	12.91	12.71	11.77	11.47	0.30	12.90	0.01
w	12.38	11.86	10.25	9.55	0.70	12.31	0.07
						Mean	0.024
						St. dev	0.030

It can be seen from Tables C.4 and C.5 that the errors (St.dev.) involved using CMC14 data and (C.2) are on average certainly less than 0.05 mag for a single star, confirming the predicted accuracy from the empirical correlation based on Landolt stars discussed earlier. If six or seven stars are used in an ensemble, then a V magnitude accuracy of ±0.03 mag is possible.

Figure C.4 shows two illustrations of the lightcurve of the 17th magnitude asteroid *2000 BD$_{19}$* depending on whether the V magnitude of comparison stars are measured directly or are calculated using data from the CMC14 catalogue. Each datapoint corresponds to a stack of 10 × 40 s unfiltered exposures using a 0.28 m aperture Schmidt–Cassegrain telescope. Note that although the images were made unfiltered, the SXV-H9 camera has its maximum response close to the V passband, such that if comparison stars are used which are similar in colour to the asteroid, the resultant differential magnitude can be transformed to the V passband. The two lightcurves depicted in Fig. C.4 are virtually identical confirming the methodology described here, and showing that the object declined in brightness from maximum to minimum in about 2 h. Subsequent (unpublished) photometry when the object was brighter showed it to have a rotational period of about 10 h.

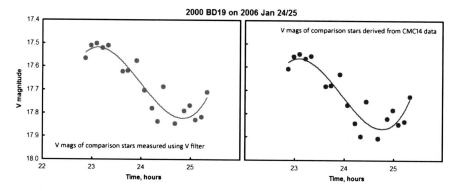

Fig. C.4. An asteroid lightcurve using standard V photometry or using CMC14-derived magnitudes

Finally, a Few Words of Caution!

Recent work has shown that observations near the Milky Way can lead to misleading results if stars suffer from a significant degree of interstellar reddening by intervening clouds of gas and dust. Imaging asteroids which lie close to the galactic plane is in any case not advisable since the field of view is often extremely crowded and it is difficult to obtain a measure of the background sky without it being contaminated by faint field stars. Similarly, as it moves the asteroid is continually encountering different field stars which also then contaminate the measuring aperture. As a general rule of thumb, avoid photometry of asteroids if they lie within a galactic latitude of +10 to −10°.

The methodology can be exported to variable star work but in this case it would be helpful to use a large FOV to include as many CMC stars as possible so that the ensemble value is then largely unaffected by possible intrinsic variability of a few of these stars. Stars fainter than about V = 15 should also be avoided to maximize photometric accuracy.

Conclusion

Equation (C.2), rounded to an adequate three significant figures, can be used to calculate V magnitudes from CMC14 data using the relationship:

$$V = 0.628 \times (J - K) + 0.995 \times r'$$

The accuracy of this correlation permits stars down to a V magnitude of 14 to be used to calculate the brightness of an asteroid to within a typical accuracy of ±0.05 mag for a single star. If an ensemble of four or more stars is used then a similar accuracy should be possible down to about V = 15. Stars located within ±10° of the galactic equator should however be avoided where possible. This approach is a very significant improvement over the use of *Tycho-2* data which has very limited application for asteroid photometry since *Tycho* can only be used down to about V = 10.5 with reasonable accuracy.

The advantage of this method is that a good number of comparison stars should be available on even a small CCD image (e.g. 12 × 8 arcmin) thus making the measurement of actual V magnitudes as simple as performing differential photometry. It has particular relevance to the determination of asteroid magnitudes, and hence absolute magnitudes, where the choice of comparison stars varies from night to night as the asteroid tracks across the sky, and the proximity to stars having well-defined magnitudes, e.g. Landolt or *Hipparcos*, is far from guaranteed.

Acknowledgements

The authors are grateful for the assistance given by Gérard Faure and John Greaves in the preparation of this paper: John for his analysis of LONEOS and 2MASS data and drawing our attention to the formula for the calculation of V magnitude, and Gérard for his encouragement in developing the methodology and suggesting its applicability to the measurement of the absolute magnitudes of asteroids. We are also indebted to the referees of this paper, David Boyd and Nick James, for their helpful and constructive comments.

Addresses: RD: 67 Haslar Crescent. Waterlooville, Hampshire
RM: Grange Cottage, Golden Hill, Stourton Caundle, Dorset

References

1. CMC14 catalogue: http://www.ast.cam.ac.uk/~dwe/SRF/camc.html
2. SDSS: http://www.sdss.org/
3. VizieR: http://webviz.u-strasbg.fr/viz-bin/VizieR
4. 2MASS: http://www.ipac.caltech.edu/2mass/
5. LONEOS photometric catalogue: ftp://ftp.lowell.edu/pub/bas/starcats/
6. Private communication, 2006 October 21
7. Landolt, A. U., *Astronomical Journal,* **104**(1), 340–371, 436–491 (1992)
8. *Hipparcos* and *Tycho* data: http://www.rssd.esa.int/index.php?project=HIPPARCOS&page=Research_tools
9. *GUIDE* software: http://www.projectpluto.com/
10. *MegaStar* software: http://www.willbell.com/software/megastar/index.htm
11. Laurie S., '*Astrometry of Near Earth Objects*', The Astronomer, 2002 June
12. Minor Planet Center's Minor Planet and Comet Ephemeris Service: http://cfa-www.harvard.edu/iau/MPEph/MPEph.html
13. *AIP4WIN* software: http://www.willbell.com/aip/index.htm
14. Astrometrica software: http://www.astrometrica.at/
15. *Aladin Sky Atlas*; http://aladin.u-strasbg.fr/aladin.gml
16. ESO On-line Digitized Sky Survey: http://arch-http.hq.eso.org/dss/dss
17. UCAC: http://ad.usno.navy.mil/ucac/
18. USNO-A2.0: http://www.nofs.navy.mil/projects/pmm/catalogs.html#usnoa
19. TASS: http://www.tass-survey.org/
20. http://www.projectpluto.com/gloss/help_18.htm#Tycho
21. AAVSO *CCD Observing Manual*: http://www.aavso.org/observing/programs/ccd/manual/
22. Boyd D., *Measuring Variable Stars using a CCD camera – a Beginner's Guide,* Brit. Astron. Assoc., 2006

Postscript

CMC14-Based Photometry Upgrade to the Astrometrica Software

In the above paper, it has been demonstrated using a few examples that the proposed empirical relationship based on Sloan-r', J and K magnitudes obtained from the CMC14 catalogue will in principle yield an accurate estimate of the V magnitude of field stars in the range, $10 < V < 15$. The relationship is especially valid for stars which are similar in colour to asteroids, so this provides a route whereby accurate V photometry of asteroids is made possible. However, for it to be a practical proposition, some degree of automation is required. Using the current version of Bill Gray's excellent planetarium program, *Guide 8.0*, CMC14 stars can be automatically downloaded via the Web enabling suitable comparison stars for any particular asteroid to be readily identified. A simple spreadsheet can then be used to calculate the V magnitudes of these stars, which with suitable photometry software enables asteroid lightcurves to be determined.

In 2008 June, following completion of the draft paper, I received a communication from Herbert Raab, the author of *Astrometrica*, saying that he was working on adding the CMC14 catalogue and he wondered whether this would be useful for photometry. *Astrometrica* (**http://www.astrometrica.at/**) is a very popular piece of software used to carry out astrometry of asteroids and comets, but the photometry it yields when based on catalogues such as the USNO B1.0 and UCAC-2 is only accurate to about 0.2 mag at best. To cut a long story short, some 10 weeks and 11 revisions of the software later, *Astrometrica* was rewritten with some help from me to permit photometry accurate to about ±0.03 mag in those regions of the sky covered by CMC14. The latest version (4.5.1.377) provides for aperture photometry in both the Johnson-V and Cousins-R passbands. The relationship described in the present paper is utilised by *Astrometrica* along with constraints on the J–K colours of reference stars, to yield V photometry to a precision of 0.01 mag. Note that the CMC14 catalogue option must be selected in the Settings configuration file for accurate photometry. Cousins-R photometry is based on the relationship, $R = r' - 0.22$.

In 2008 September, the binary asteroid 2000 DP_{107} made a close approach to the Earth and I was able to secure a series of about 500×30 s exposures on the night of September 26/27 from Golden Hill Observatory, Dorset (MPC Code J77). This activity is in support of the Ondrejov Observatory Survey of Binary Asteroids, which is led by Dr Petr Pravec.

During the observing run, the 15th magnitude object traversed some three different fields of view. The image frames were processed using *Astrometrica*, with dark frames subtracted and flat-fields applied to each frame in turn before identifying an ensemble of suitable stars, which are then used *en masse* to determine the V magnitude and exact position of the moving asteroid. The resultant lightcurve comprising 462 datapoints is depicted in Fig. C.5.

It can be seen that a very acceptable result has been obtained given the faintness of the object and the size of the telescope used (28 cm aperture). The irregular nature of the lightcurve is a consequence of it being a binary system, comprising a primary body some 800 m across which rotates every 2.78 h, around which orbits a secondary object about 350 m in size, completing a single revolution every 42.2 h. The secondary is thought to be locked in synchronous rotation about the primary (rather like the Earth–Moon system) and the system occasionally undergoes

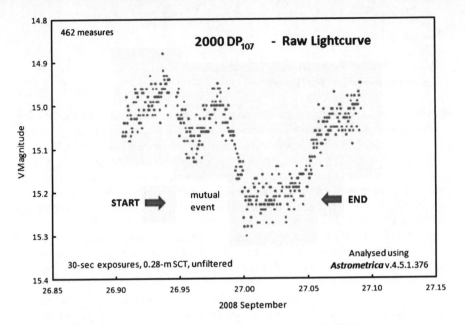

Fig. C.5. Raw lightcurve of the binary asteroid 2000 DP₁₀₇ produced using *Astrometrica* upgraded for CMC14 photometry

mutual eclipses and occultations. One such event, lasting about 2 h, happened to take place during my observing run of September 26/27: the start and end of the event have been highlighted in Fig. C.5. Since the form of the lightcurve outside of eclipse/occultation is relatively complex, this has been subtracted by Petr Pravec from the raw lightcurve to produce Fig. C.6, which depicts the contribution arising from the mutual event alone.

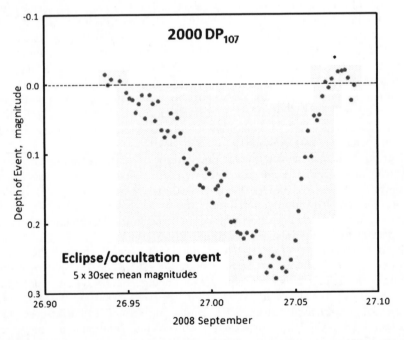

Fig. C.6. Lightcurve showing details of an eclipse/occultation event, obtained after subtracting the normal rotational lightcurve of the binary system from the data shown in Fig. C.5

Given that the target was 15th magnitude at the time, the form of the event is very well-defined and clearly demonstrates success in upgrading *Astrometrica* for photometry of asteroids. It should be noted that in the majority of cases, the use of two different configuration file settings in *Astrometrica* is preferred: one optimised for astrometry and one for photometry (see 'Notes on using Astrometrica to determine V magnitudes' below). This is especially true if the region of sky traversed by the asteroid is very crowded with field stars.

I would like to express my sincerest gratitude to Herbert Raab for his patience and perseverence in making the necessary modifications to his software – I am most grateful to him.

Richard Miles

Director, Asteroids and Remote Planets Section, British Astronomical Association

November 14, 2008

Notes on Using *Astrometrica* to Determine V Magnitudes

Astrometrica, versions 4.5.0.376 or later, incorporates CMC14 data which allows V magnitudes to be determined. This version is a major advance in that, whether the asteroid is a Main Belt Asteroid or a fast-moving Near Earth Object crossing many fields of view, the software is able to yield accurate magnitudes as well as positions with the minimum of effort. It should be noted however that the Minor Planet Center recommend, in their 'Guide to Minor Planet Astrometry', the use of the USNO-B1.0 catalogue for astrometry. The CMC14 catalogue does not include proper motions and therefore its accuracy will degrade as we move away from the epoch of the catalogue positions. Images must therefore be analysed using the USNO-B1.option for astrometry and then again using the CMC14 option for photometry.

MPC Reporting

MPC reports only allow for one catalogue to be identified therefore a suggested format, mentioning the use of the CMC14 catalogue, by the use of 'COM' (comment) lines is:

```
COD G68
COM Photometry uses CMC14 catalogue
COM transformed to V using formula
COM V = 0.628*(J–K) + 0.995*r'
OBS R.Dymock
MEA R.Dymock
TEL 0.61-m f/10 reflector + CCD
ACK MPCReport file updated 2009.05.28 16:59:28
AC2 roger.dymock@ntlworld.com
NET USNO-B1.0
```

01909	C2009 05 27.21539 11 56 16.43 -00 34 02.8	16.0 V	G68
01909	C2009 05 27.26725 11 56 18.79 -00 34 13.7	15.9 V	G68
01909	C2009 05 27.31910 11 56 21.10 -00 34 24.4	15.9 V	G68

----- end -----

Photometry Settings

CCD tab:

Choose the V or R passband appropriate to the filter used

Set Saturation to 50,000 to ensure any stars which saturate pixels are not included in the solution

If working in unfiltered mode, either 'V' or 'R' magnitudes can be reported provided that the colour response of your unfiltered CCD camera is close to the chosen option. For most astronomical CCD cameras, the 'R' magnitude option is best although Sony interline transfer chips are closest to 'V' in their response. You can experiment with both of these options by checking the results in terms of the residuals reported in the Log file before deciding which is better for your camera.

Program tab:

Under Star Catalog select CMC-14,

Set the magnitude Lower Limit to 14.5 or even 14 (rather than a fainter limit) as this can further improve the accuracy of the photometry,

Under 'MPC Report', select 'Magnitude to 0.1 mag',

Under 'Object Detection' the size of the 'Aperture Radius' selected also equals the size of the aperture used to perform photometry on objects on the frame. Normally, users should adjust the value of the 'Aperture Radius' so that it is large enough to contain the entire visible image of each comparison star and the asteroid,

Under 'Object Detection, Background from' select 'Aperture' (PSF must not be used for photometry)

Under 'Residuals' you can set the 'Photometric Limit' as low as 0.20 mag without rejecting a large fraction of potential reference stars.

So long as you have at least say 6–8 stars on each image, the photometry should be accurate to better than 0.05 mag (provided the signal–to–noise of the asteroid is adequate). To improve signal to noise ratio use the Track and Stack facility to co-add a number of image frames by keying in the speed and direction of motion of the asteroid, or selecting the asteroid from the drop down list, and choosing 'Average' for the final stacked image. The ultimate accuracy of the frame-to-frame calibration (zeropoint) depends on the availability of reference stars but can easily attain 0.02 or even 0.01 mag. Fortunately, the CMC14 catalogue numbers over 95 million stars of which about 60 million are suitable for use as reference stars and so it is usual to have quite a number of useable CMC14 stars in any one frame.

Astrometry Settings

The only option that needs to be changed is to select the USNO-B 1.0 catalogue in Star Catalog under the Program tab.

Richard Miles and Roger Dymock

Appendix D

Astrometry How-To

Dear Observers,

We all too frequently receive error-filled submissions of astrometry here at the Minor Planet Center (MPC), and I've decided to write a little "how-to" describing what common pitfalls can be avoided, and how to improve, in general, what folks submit to the MPC.

First, please read the *Guide to Minor Body Astrometry* at: http://www.minorplanetcenter.org/iau/info/Astrometry.html. That document and this 'how-to' overlap significantly, so please read both carefully.

Most Common Problems

False detections due to hot pixels, bad pixels, poor flat-fielding, and internal reflections

These are the most common mistakes made in data submitted to the MPC. It isn't just amateurs who make this mistake. Professional observers still report false detections of this type every week. The key is to avoid lining up these objects and "creating" a real object. By far the easiest thing to do is simply dither the telescope between successive images. This will eliminate 99% of all false detections immediately. At CSS we had this problem, and we eventually settled on simply moving the telescope slightly before each exposure. An example pattern is as follows: image 1 was at the expected coordinates, image 2 was 30 arcsec north of this location, image 3 was 30 arcsec south of the expected coordinates, and image 4 was 30 arcsec south and west of the expected coordinates. This simple procedure makes it nearly impossible for anything fake to line up with linear motion within a few arcsecs tolerance.

The second problem with false detections is using a bare minimum of images. I would never use two images for discovery purposes unless I had very small pixels (0.5"/ pixel or so), and I had a good point-spread function for both detections. Three images is, in my opinion, the bare minimum for consideration. However I do not favor this method. Simply use more than three images, and dither the telescope for best results.

Another very common mistake is absolute bare-minimum time intervals between images. It is best to have at least 30 m worth of coverage on each and every object. Note that CSS, Lincoln Near Earth Asteroid Research (LINEAR), and Spacewatch all have $t > 30$ m. LINEAR averages around 70 m between images on each object. This gives a more robust Vaisala orbit, better linking probabilities on the subsequent nights, and lastly, prevents bad links by the observer and the MPC. These short-interval links are often misidentified, or worse, spurious objects! The MPC now requires $t > 20$ m for designations except in extreme circumstances.

Keep in mind this is t > 20 m on each night. In addition, almost all false detections won't show linear motion over a 20 or 30 m interval, but can show nearly linear motion on shorter timescales.

Obviously Bad Astrometry

It is surprising that this is the second most common problem, and perhaps the most annoying. We often get observations of objects that are so clearly wrong, a simple cursory examination would show this. For example, we still receive reports every few days of objects that are reversing direction of motion between three measurements, or worse, two good positions and one that is off by 5 arcmin or more.

Also in this category are bad links by the observer, which happen on a weekly basis. For example, an observer will go to the expected coordinates for an object on the MPC's Near Earth Object Confirmation Page (NEOCP), find a bright object, measure it, and send it in as the NEOCP object only to find later this was a routine numbered asteroid. This is precisely why the NEOCP, and the MPC Ephemeris Service, provide you with the speed and direction of the object in question. I'm sure you'd be surprised how many routine MBAs moving 30 arcsec/h are turned in where the observer thought this was an NEOCP object that was supposed to be moving 300 arcsec/h. We also receive bad links where observers simply went to the MPChecker, found the object closest to their object, and pasted this designation in the observation string. This causes me no end of grief, because in most cases, it is simply easier to paste a new observer-assigned temporary designation on each object submitted. In this fashion all "new" objects are identified by our automatic software and will receive new provisional designations which are e-mailed automatically. This also reduces e-mail traffic back and forth between parties (one thing that definitely helps me, since I get a few hundred e-mails per day).

Time Problems

Of all things, this should never be a problem, but it occurs frequently. It is imperative that the observer check the clock each and every time observations are taken. Badly timed images are not only an amateur problem – every single currently active professional survey has had some sort of timing problem, from bad local time, UT time corrections, bad computer clock time, mistiming the exposure start and end, and miscalculating the midpoint of the exposure. Yes, these were all done in one form or another by professional surveys. Please do be careful here as there's no excuse for this error and trust me, I'm talking from experience. At one time CSS had a minus 12 s error on all images due to improper coding of the start time in the FITS header, and this was more or less my fault for not checking it!

Junk Astrometric Solutions

Another frequent problem is astrometry that is clearly the right object but is also clearly incorrect. This arises in some cases for horrible or non-converged fits on astrometric solutions. If your RMS on your solution is extremely small and uses only three or four stars, it is probably wrong. Likewise, if it is over about 0.7 arcsec, you've also probably got a problem – keep a close eye on those solutions!

Over-Observing Bright Objects (Not an Astrometric Problem, But Something of Note)

A sizable number of objects need no astrometry whatsoever, and yet we still will receive literally thousands of observations each month of these objects. Bright numbered asteroids, unless they are occultation targets, radar targets, or mission targets, really don't need astrometry, so don't go out of your way to target them. You may, and should, measure them if they just happen to appear in your frames, but targeted astrometric observations of routine numbered MBAs by amateur astronomers are almost certainly a waste of time. Likewise, many non-numbered NEOs are absolutely hammered by amateurs for no apparent reason. A good rule of thumb is to only observe objects that you and your system can actually help. So if the current ephemeris uncertainty is only 0.3 arcsec, there is no possible way you can dramatically improve the orbit if your astrometry is only good to 0.5 arcsec. Also, given the sky coverage and sheer number of professional surveys in action, it is very likely that your hard work will simply be obviated the next night by a survey.

Single Positions

Please cease and desist from sending single positions of any object on a given night, unless this object is spectacularly important, and then only do so with gratuitous comments regarding the accuracy of the measure and why no other measures were obtained. For example, this might be acceptable for NEOs and comets from skilled observers, but single, isolated positions for MBAs are very likely to be deleted.

Notes Regarding Professional Programs

At this point, it is probably important for amateurs to know a thing or two about the professional programs. LINEAR takes five images of each field, total spacing about 1 h. CSS, Siding Spring Survey (SSS), and Lowell Observatory Near Earth Object Search (LONEOS) all take four images of each field, with intervals varying from 20 to 60 m from first to last image. Spacewatch and Near Earth Asteroid Tracking (NEAT) take only three images, spaced by 20–60 m, but they both have small (~1 arcsec) pixels. Each and every one of the aforementioned programs submits all objects as new objects, meaning each object observed on a given night has its own, observer-assigned unique temporary designation. These observations pass flawlessly through our automatic processing code, and the remaining one-night objects that might be NEOs are left for further examination. Given that a good deal of MPC effort has been put forth to process the bulk of the data this way, other observers should consider operating in a like manner.

Tim Spahr
Director, Minor Planet Center
Former observer with the Catalina Sky Survey (CSS)

Index

Printed by Publishers' Graphics LLC
DBT140408.15.17.8 20140408